Designerly Ways of Knowing

Nigel Cross

Designerly Ways of Knowing

With 15 Figures

 Springer

Nigel Cross, PhD
Professor of Design Studies
Department of Design and Innovation
Faculty of Technology
The Open University
Milton Keynes
MK7 6AA
UK

British Library Cataloguing in Publication Data
Cross, Nigel, 1942-
   Designerly ways of knowing
   1.Design, Industrial - Methodology 2.Creation (Literary,
   artistic, etc.) 3.Creative ability in technology
   I.Title
   745.2
ISBN-13: 9781846283000
ISBN-10: 1846283000

Library of Congress Control Number: 2006921168

ISBN-10: 1-84628-300-0   e-ISBN 1-84628-301-9   Printed on acid-free paper
ISBN-13: 978-1-84628-300-0

Printed in Germany

9 8 7 6 5 4 3 2 1

Springer Science+Business Media
springer.com

Preface

This book traces the development of a personal research programme over a period of many years. The starting point for the programme was a realisation that research in design seemed to have no clear goal of what it was trying to achieve. A key insight for me was to realise that if we wanted to develop a robust, independent discipline of design (rather than let design be subsumed within paradigms of science or the arts), then we had to be much more articulate about the particular nature of design activity, design behaviour and design cognition. We had to build a network of arguments and evidence for 'designerly ways of knowing'.

The research programme has included some empirical, laboratory-based work, but has also included theoretical reflection, and attempts to review and synthesise the work of other researchers. I have reported this work at various times and in various places – in lectures, conference presentations and journal papers. In this book I have brought together a selected series of these reports, trying to trace a coherent thread, and to lay out some of the network of arguments and evidence referred to above. My goal has been to understand how designers think, or the nature of design expertise, trying to establish its particular strengths and weaknesses, and giving credit where it might be due for design cognition as an essential aspect of human intelligence. The versions of the reports and papers now published here have been revised in order to aid coherence and avoid overlap.

Chapter 1, 'Designerly Ways of Knowing' was first published as a contribution to a series on 'Design as a Discipline' in the journal *Design Studies*, which aimed to establish the theoretical bases for treating design as a coherent discipline of study. The first contribution in the series had been from Bruce Archer, in the very first issue of *Design Studies* in 1979, in which he outlined arguments for a 'third area' of education – design. At that time, the subject of 'design' was being introduced into secondary schools in the UK for the first time, meaning that it was becoming part of general, not specialist education. Also at that time, my colleagues and I at the new Open University had been facing similar issues concerned with how to develop a form of design education that was relevant for 'everyone', and was not necessarily aimed at preparing students for a professional role in design practice. In my contribution to the debate I attempted to develop further an understanding of this 'third area' by contrasting it with the other two – sciences and humanities – and to go on to consider the criteria which design must satisfy to be acceptable as a part of general education. I argued that such an acceptance must imply a reorientation from the instrumental aims of conventional, professionally-orientated design education, towards intrinsic values of design as a valid subject of study for everyone. These intrinsic values, I suggested, must derive from the deep, underlying patterns of how designers think and act, or the 'designerly ways of knowing'. Because of their common concern with these fundamental 'ways of knowing', I suggested that both design research and design education were thereby contributing to the development of design as a discipline. I also suggested that this emerging view that 'there are designerly ways of knowing' could form the axiomatic 'touch-stone theory' for research within design as a discipline.

The second chapter, 'The Nature and Nurture of Design Ability', is based on my inaugural lecture as Professor of Design Studies in the Open University, which I delivered in 1989. The first part of my lecture concentrated on the nature of design ability, for which I drew upon a variety of studies and investigations into design activity and designer behaviour. From a review of these studies, I summarised design ability as comprising abilities of resolving ill-defined problems, adopting solution-focused cognitive strategies, employing abductive or appositional thinking and using non-verbal modelling media. These abilities are highly developed in skilled designers, but I suggested that they are also possessed to some degree by everyone. I then outlined a case for design ability as a fundamental form of human intelligence, thus seeking to provide a much broader foundation for establishing 'designerly ways of knowing'. In the second part of my lecture I argued that understanding the nature of design ability is necessary in order to enable design educators to nurture its development in their students. I discussed the nurture of this ability through design education, with particular reference to the problem of providing design education through the distance-learning media of the Open University. In the chapter here, I have revised the second part of the lecture so as to make it less focused on the special concerns of the Open University, and hopefully more relevant to all design educators, although still emphasising that 'open-ness' is a key principle for modern design education.

Chapter 3, 'Natural and Artificial Intelligence in Design', arose from the challenge of being asked to give the keynote lecture to the international conference on Artificial Intelligence in Design, in 1998. Rather than discussing 'artificial intelligence', I chose to concentrate on the 'natural intelligence' of design. The lecture addressed what we know about the 'natural intelligence' of design ability, and the nature of design activity. My starting point was the observation that the ability to design is widespread amongst all people, but some people appear to be better designers than others. I used quotations and comments from some acknowledged expert designers to reinforce general findings about the nature of design activity that have come from design research. I also referred to and analysed the role of sketching in design in order to exemplify some of the complexity of designing. Finally, I made some comments about the value and relevance of research into artificial intelligence (AI) in design. I suggested that one aim of research in AI in design should be to help inform understanding of the natural intelligence of design ability, to help us better to understand such natural intelligence, or human cognition.

A key aspect of human cognition in the context of design is creative thinking. The next two chapters report some studies of creative cognition in design.

In Chapter 4, 'The Creative Leap', I report an attempt to analyse how creative thinking happens in design, through one case study which was captured in a protocol study experiment. The 'creative leap', in which a novel concept emerges – perhaps quite suddenly – as a potential design solution, is widely regarded as a characteristic feature of creative design. The investigation reported here is based on an example of a 'creative leap' which occurred during a recorded laboratory study of the activity of a small design team. The characteristics and context of this 'creative leap' are reconstructed from the recorded material, and I use the commonly accepted procedures underlying generic descriptive models of creative design to try to provide further insight into the example. I also make some observations of potential implications for computer modelling of creative design. I conclude that the perceptual act underlying creative insight in design is not so much a 'leap', but more akin to 'bridging' between problem space and solution space. This fits with the appositional nature of design thinking, in that the bridging concept embodies satisfactory relationships between problem and solution.

Chapter 5, 'Creative Strategies', continues the line of investigation into creative cognition in design. Three studies of innovative design in engineering and product design are reported. As well as the small team design project reported in Chapter 4, my colleagues and I had been able to capture in the same series of experiments the think-aloud protocols of an outstanding engineering designer, Victor Scheinman. I have also been fortunate enough to be able to conduct some in-depth interviews with a couple of other truly outstanding designers – the product designer Kenneth Grange and the Formula One racing car designer Gordon Murray. In this chapter I present an example of each of these outstanding designers' approaches to a particular design problem that they faced. I have tried to draw comparisons between the three examples, and there do appear to be some striking similarities in their adoption of a strategic design approach, despite the very different project examples. I develop a general descriptive model from the examples, showing how strategic knowledge in creative design is exercised at three levels: low level articulated knowledge of first principles, an intermediate level of tacit personal and situated knowledge applied within the particular problem and its context, and high level implicit and explicit knowledge of problem goals and criteria. All three outstanding designers seem to exercise this strategic knowledge in similar ways in creating novel design proposals.

The studies of Victor Scheinman and the three-person design team in the previous two chapters drew upon the experimental method of protocol analysis, which has become the most widely used technique for investigating design cognition. Chapter 6, 'Understanding Design Cognition', reviews a number of protocol and other such empirical studies of design activity, and summarises results relevant to understanding the nature of design cognition from an interdisciplinary, domain-independent overview. The results are presented grouped into three major aspects of design cognition – the formulation of problems, the generation of solutions, and the utilisation of design process strategies. I draw parallels and comparisons between results, and I find many similarities of design cognition across domains of professional practice. Perhaps the most interesting conclusion is that it seems that the 'intuitive' behaviour of experienced designers is often highly appropriate to the special nature of design tasks, although appearing to be 'unprincipled' in theory.

The final chapter returns to the starting theme of 'Design as a Discipline'. It begins by unravelling some of the history of concern with relationships between design and science. In the original conference paper I was seeking to develop a view of design as a discipline based upon a science of design, but not a 'design science'. As in Chapter 1, I argue that the underlying axiom of this discipline is that there are forms of knowledge peculiar to the awareness and ability of a designer. In the latter part of the chapter I outline the ways in which this discipline of design, and the understanding of designerly ways of knowing, can be pursued through design research. I identify three sources of design knowledge as research loci: people, processes and products. These are the foundation stones for understanding designerly ways of knowing.

In selecting this particular set of speeches, papers and reports, I have tried to construct some of the argument, and to assemble some of the evidence that supports the concept of 'designerly ways of knowing'. I believe that this concept can now be justified, and that – thanks to the work of several other design researchers besides myself – we now have a much clearer view of what constitutes the particular nature of design cognition.

The span of time covered by the papers that formed the original versions of the seven chapters here is some twenty years. One might have hoped that much more would have been achieved over such a long period. But the discipline of design is still quite young, and it still has a relatively small research base. I hope that this book might serve to record the coming of age of the discipline, to cement the discipline's foundations, and to suggest ways forward for a new and rapidly growing generation of design researchers.

Nigel Cross

Acknowledgements

My wife Anita has also been my working partner in much of the research reported here, and she has been a constant source of inspiration and support.

The Open University has provided me with time and resources to undertake personal research, and with study leave in order to complete this book.

Chapter 4 and part of Chapter 5 are based on data from the Delft Protocols Workshop, 1994, organised by Kees Dorst, Henri Christiaans and myself at Delft University of Technology, in association with Steve Harrison and Scott Minneman of Xerox Palo Alto Research Center. The Workshop was made possible by financial and practical support provided by the Faculty of Industrial Design Engineering of Delft University of Technology, Xerox PARC and the Engineering Design Center of Stanford University. Above all, gratitude is due to Victor Scheinman and the other designers who willingly participated in the experiments, provided their time and talent free of charge, and allowed their design activity to be observed and analysed.

Kenneth Grange and Gordon Murray generously provided time to discuss their work, reported in Chapter 5.

I am grateful to the following for permission to republish papers: to Elsevier for Chapters 1, 2 and 3 from *Design Studies*; to the Key Centre of Design Computing and Cognition, University of Sydney, for Chapter 4 from *Computational Models of Creative Design III* and Chapter 5 from *Strategic Knowledge and Concept Formation III*; to Elsevier for Chapter 6 from *Design Knowing and Learning*; to the Politecnico di Milano for Chapter 7 from *Design+Research*.

Grateful acknowledgement is made to the following sources of illustrations: Figures 3.1, 3.2 Iain Fraser and Rod Henmi, *Envisioning Architecture*, John Wiley & Sons Inc., New York; Figure 4.2 Joachim Gunter, Eckart Frankenberger and Peter Auer, Figure 4.3 Maryliza Mazijoglou, Stephen Scrivener and Sean Clark, Figure 4.4 David Radcliffe, Figure 4.5 Gabriela Goldschmidt, all from *Analysing Design Activity* (eds. N Cross, H Christiaans and K Dorst), John Wiley & Sons Ltd., Chichester, reproduced with permission.

Contents

1

Designerly Ways of Knowing

A principal outcome of a research project at the Royal College of Art on 'Design in General Education' was the statement of a belief in a missing 'third area' of education. The two already-established areas can be broadly classified as education in the sciences and education in the arts, or humanities. These 'two cultures' have long been recognised as dominating our social, cultural and educational systems. In the traditional English educational system, especially, children have been required to choose one or other of these two cultures to specialise in at a relatively early age.

The 'third culture' is not so easily recognised, simply because it *has* been neglected, and has not been adequately named or articulated. In their report (Royal College of Art, 1979), Bruce Archer and his colleagues were prepared to call it 'Design with a capital D' and to articulate it as 'the collected experience of the material culture, and the collected body of experience, skill and understanding embodied in the arts of planning, inventing, making and doing'.

From the RCA report, the following conclusions can be drawn on the nature of 'Design with a capital D':

- The central concern of Design is 'the conception and realisation of new things'.
- It encompasses the appreciation of 'material culture' and the application of 'the arts of planning, inventing, making and doing'.
- At its core is the 'language' of 'modelling'; it is possible to develop students' aptitudes in this 'language', equivalent to aptitudes in the 'language' of the sciences (numeracy) and the 'language' of humanities (literacy).
- Design has its own distinct 'things to know, ways of knowing them, and ways of finding out about them'.

First published in *Design Studies* Vol 3, No 4, October 1982, pp. 221–227.

Even a 'three cultures' view of human knowledge and ability is a simple model. However, contrasting design with the sciences and the humanities is a useful, if crude, way of beginning to be more articulate about it. Education in any of these 'cultures' entails the following three aspects:

- the transmission of knowledge about a phenomenon of study
- a training in the appropriate methods of enquiry
- an initiation into the belief systems and values of the culture

If we contrast the sciences, the humanities, and design under each aspect, we may become clearer of what we mean by design, and what is particular to it.

- The phenomenon of study in each culture is
  - o in the sciences: the natural world
  - o in the humanities: human experience
  - o in design: the artificial world

- The appropriate methods in each culture are
  - o in the sciences: controlled experiment, classification, analysis
  - o in the humanities: analogy, metaphor, evaluation
  - o in design: modelling, pattern-formation, synthesis

- The values of each culture are
  - o in the sciences: objectivity, rationality, neutrality, and a concern for 'truth'
  - o in the humanities: subjectivity, imagination, commitment, and a concern for 'justice'
  - o in design: practicality, ingenuity, empathy, and a concern for 'appropriateness'

In most cases, it is easier to contrast the sciences and the humanities (*e.g.* objectivity *versus* subjectivity, experiment *versus* analogy) than it is to identify the relevant comparable concepts in design. This is perhaps an indication of the paucity of our language and concepts in the 'third culture', rather than any acknowledgement that it does not really exist in its own right. But we are certainly faced with the problem of being more articulate about what it means to be 'designerly' rather than to be 'scientific' or 'artistic'.

Perhaps it would be better to regard the 'third culture' as technology, rather than design. This 'material culture' of design is, after all, the culture of the technologist – of the designer, doer and maker. Technology involves a synthesis of knowledge and skills from both the sciences and the humanities, in the pursuit of practical tasks; it is not simply 'applied science', but 'the application of scientific and *other organised knowledge* to practical tasks...' (Cross, *et al.,* 1981).

The 'third culture' has traditionally been identified with technology. For example, the philosopher A N Whitehead (1932) suggested that: 'There are three main roads along which we can proceed with good hope of advancing towards the best balance of intellect and character: these are the way of literary culture, the way of scientific culture, the way of technical culture. No one of these methods can be exclusively followed without grave loss of intellectual activity and of character.'

Design in General Education

I think it is no accident that a fundamental reconceptualising of design emerged from a project, such as the Royal College of Art's, related to the development of design in general education. Our established concepts of design have always been related to specialist education: design education has been preparation of students for a professional, technical role. But now we are exploring the ways and the implications of design being a part of everyone's education, in the same ways that the sciences and the humanities are parts of everyone's education.

Traditionally, design teachers have been practicing designers who pass on their knowledge, skills and values through a process of apprenticeship. Design students 'act out' the role of designer in small projects, and are tutored in the process by more experienced designers. These design teachers tend to be firstly designers, and only secondly and incidentally teachers. This model may be defensible for specialist education, but in general education all teachers are (or should be) firstly teachers, and only secondly, if at all, specialists in any field.

To understand this distinction we must understand the differences between specialist education and general education. The main distinction lies in the difference between the instrumental, or *extrinsic,* aims that specialist education usually has, and the *intrinsic* aims that general education must have. It is perfectly acceptable for architectural education, say, to have the instrumental aim of providing competent designers of buildings, but this cannot be an aim of general education. Anita Cross (1980) has pointed out that, 'Since general education is *in principle* non-technical and non-vocational, design can only achieve parity with other disciplines in general education if it is organised as an area of study which contributes as much to the individual's self-realisation as to preparation for social roles.'

Whatever government ministers or industrialists may think, the aim of general education is not the preparation of people for social roles. In a sense there is no 'aim' to general education. The educationist Peters (1965) claims that:

> It is as absurd to ask what the aim of education is as it is to ask what the aim of morality is... The only answer that can be given is to point to something intrinsic to education that is regarded as valuable such as the training of intellect or character. For to call something 'educational' is to intimate that the processes and activities themselves contribute to or involve something that is worthwhile... People think that education must be for the sake of something extrinsic that is worthwhile, whereas the truth is that being worthwhile is part of what is meant by calling it 'education'.

Educational Criteria

According to Peters the concept of education is one which only suggests *criteria* by which various activities and processes can be judged to see if they can be classified as 'educational'. Thus, giving a lecture *may* be educational, but it might not be if it does not satisfy the criteria; a student design project *may* be educational, but also might not be.

Peters suggests three principal criteria for education, the first of which is that worthwhile knowledge of some value must be transmitted. This first criterion seems straightforward, but actually raises problems of defining what is 'worthwhile'. The example offered by Peters is simplistic: 'We may be educating someone while we are training him: but we need not be. For we may be training him in the art of torture.' Deciding what is worthwhile is obviously value-laden and problematic. We might all agree that 'the art of torture' hardly counts as worthwhile, but what about, say, 'the art of pugilistics'? However, 'the arts of planning, inventing, making and doing' (to draw on Archer's definition of design again) are presumably clearly recognised as 'worthwhile'.

Peters' second criterion derives from his concern with the processes by which students are educated. He stresses that the *manner* in which people are educated is just as important as the *matter* which is transmitted:

Although 'education' picks out no specific processes it does imply criteria which processes involved must satisfy in addition to the demand that something valuable must be passed on. It implies, first of all, that the individual who is educated shall come to care about the valuable things involved, that he shall want to achieve the relevant standards. We would not call a man 'educated' who knew about science but cared nothing for truth or who regarded it merely as a means to getting hot water and hot dogs. Furthermore it implies that he is initiated into the content of the activity or forms of knowledge in a meaningful way, so that he knows what he is doing. A man might be conditioned to avoid dogs or induced to do something by hypnotic suggestion. But we could not describe this as 'education' if he did not know what he was learning while he learned it.

This second criterion of 'education' therefore stresses the need for the student to be both self-aware and aware of what and why he is learning. It is a process neither of imposing patterns on the student's mind, nor of assuming that free growth towards a desirable end will somehow occur without guidance. Education must be designed deliberately to enhance and to develop students' intrinsic cognitive processes and abilities.

Peters' third criterion derives from the consideration that: 'We often say of a man that he is highly trained, but not educated. What lies behind this condemnation? ... It is ... that he has a very limited conception of what he is doing. He does not see its connection with anything else, its place in a coherent pattern of life. It is, for him, an activity which is cognitively adrift.'

Peters concludes from this consideration that 'education' is related to 'cognitive perspective', which 'explains why it is that some activities rather than others seem so obviously to be of educational importance. There is very little to know about riding bicycles, swimming, or golf. It is largely a matter of "knowing how" rather than of "knowing that" (Ryle, 1949) – of knack rather than understanding. Furthermore what there is to know throws very little light on much else.'

This is therefore a challenging criterion for design education, since design is often regarded as a skill, perhaps something like bicycle-riding, swimming, or playing golf. Indeed, elsewhere we have used Ryle's distinction between 'knowing how' and 'knowing that' to emphasise the role of 'know how' in design (Cross, *et al.*, 1981). However, I would now accept Peters' suggestion that:

> An 'educated man' is distinguished not as much by what he does as by what he 'sees' or 'grasps'. If he does something very well, in which he has to be trained, he must see this in perspective, as related to other things. It is difficult to conceive of a training that would result in an 'educated' man in which a modicum of instruction has no place. For being educated involves 'knowing that' as well as 'knowing how'.

So to satisfy this third criterion of 'education', simple training in a skill is not enough. One *is* 'trained' as a designer, or doctor, or philosopher, but that alone does not make one 'educated'.

I have considered Peters' three criteria for 'education' at some length because it is important for the proponents of design in general education to be able to meet such criteria. It entails a fundamental change of perspective from that of a vocational training for a design profession, which is the only kind of 'design education' we have had previously. Design in general education is *not* primarily a preparation for a career, *nor* is it primarily a training in useful productive skills for 'doing and making' in industry. It must be defined in terms of the *intrinsic* values of education.

The interpretation of 'education' that Peters has developed, then, stresses its intrinsic merits. To be educated is of value in and of itself, not because of any extrinsic motivating factors or advantages it might be considered to offer, such as getting a job. In order to justify design as a part of general education, therefore, it is necessary to ensure that what is learned in design classes, and the way it is learned, can meet these criteria. We have to be able to identify that which is intrinsically valuable in the field of design, such that it is justifiably a part of everyone's education and contributes to the development of an 'educated' person.

Ways of Knowing in Design

The claim from the Royal College of Art study of 'Design in General Education' was that 'there are things to know, ways of knowing them, and ways of finding out about them' that are specific to the design area. The authors imply that there are designerly ways of knowing, distinct from the more usually-recognised scientific and scholarly ways of knowing. However, the Royal College of Art authors do little to explicate these designerly ways of knowing. They do point out that 'it would not do to accept design as a sort of ragbag of all the things that science and the humanities happen to leave out,' but they are less than precise about what design should include. Design must have its own inner coherence, in the ways that science and the humanities do, if it is to be established in comparable intellectual and educational terms. But the world of design has been badly served by its intellectual leaders, who have failed to develop their subject *in its own terms*. Too often, they have been seduced by the lure of *Wissenschaft*, and turned away from the lore of *Technik*; they have defected to the cultures of scientific and scholarly enquiry, instead of developing the culture of designerly enquiry.

So what can be said about these ill-defined 'ways of knowing' in design? There has, in fact, been a small and slowly-growing field of enquiry in design research from which it is possible to begin to draw some conclusions.

Design Processes

For example, a number of observational studies has been made of how designers work. These studies tend to support the view that there is a distinct 'designerly' form of activity that separates it from typical scientific and scholarly activities. Lawson's (1979) studies of design behaviour, in particular, have compared the problem-solving strategies of designers with those of scientists. He devised problems which required the arrangement of 3D coloured blocks so as to satisfy certain rules (some of which were not initially disclosed), and set the same problems to both postgraduate architectural students and postgraduate science students. The two groups showed dissimilar problem-solving strategies, according to Lawson. The scientists generally adopted a strategy of systematically exploring the possible combinations of blocks, in order to discover the fundamental rule which would allow a permissible combination. The architects were more inclined to propose a series of solutions, and to have these solutions eliminated, until they found an acceptable one. Lawson (1980) has commented:

> *The essential difference between these two strategies is that while the scientists focused their attention on discovering the rule, the architects were obsessed with achieving the desired result. The scientists adopted a generally problem-focused strategy and the architects a solution-focused strategy. Although it would be quite possible using the architect's approach to achieve the best solution without actually discovering the complete range of acceptable solutions, in fact most architects discovered something about the rule governing the allowed combination of blocks. In other words, they learn about the nature of the problem largely as a result of trying out solutions, whereas the scientists set out specifically to study the problem.*

These experiments suggest that scientists problem-solve by analysis, whereas designers problem-solve by synthesis. Lawson repeated his experiments with younger students and found that first-year students and sixth-form school students could not be distinguished as 'architects' and 'non-architects' by their problem-solving strategies: there were no consistent differences. This suggests that architects learn to adopt their solution-focused strategy during, and presumably as a result of, their education. Presumably, they learn, are taught, or discover, that this is the more effective way of tackling the problems they are set.

A central feature of design activity, then, is its reliance on generating fairly quickly a satisfactory solution, rather than on any prolonged analysis of the problem. In Simon's (1969) inelegant term, it is a process of 'satisficing' rather than optimising; producing any one of what might well be a large range of satisfactory solutions rather than attempting to generate the one hypothetically-optimum solution. This strategy has been observed in other studies of design behaviour, including engineers (Marples, 1960), urban designers (Levin, 1966) and architects (Eastman, 1970).

Why it should be such a recognisably 'designerly' way of proceeding is probably not just an embodiment of any intrinsic inadequacies of designers and their education, but is more likely to be a reflection of the nature of the design task and of the nature of the kinds of problems designers tackle. The designer is constrained to produce a practicable result within a specific time limit, whereas the scientist and scholar are both able, and often required, to suspend their judgements and decisions until more is known – 'further research is needed' is always a justifiable conclusion for them.

It is also now widely recognised that design problems are ill-defined, ill-structured, or 'wicked' (Rittel and Webber, 1973). They are not the same as the 'puzzles' that scientists, mathematicians and other scholars set themselves. They are not problems for which all the necessary information is, or ever can be, available to the problem-solver. They are therefore not susceptible to exhaustive analysis, and there can never be a guarantee that 'correct' solutions can be found for them. In this context a solution-focused strategy is clearly preferable to a problem-focused one: it will always be possible to go on analysing 'the problem', but the designer's task is to produce 'the solution'. It is only in terms of a conjectured solution that the problem can be contained within manageable bounds (Hillier and Leaman, 1974). What designers tend to do, therefore, is to seek, or impose a 'primary generator' (Darke, 1979) which both defines the limits of the problem and suggests the nature of its possible solution.

In order to cope with ill-defined problems, designers have to learn to have the self-confidence to define, redefine and change the problem-as-given in the light of the solution that emerges from their minds and hands. People who seek the certainty of externally structured, well-defined problems will never appreciate the delight of being a designer. Jones (1970) has commented that 'changing the problem in order to find a solution is the most challenging and difficult part of designing'. He also points out that 'designing should not be confused with art, with science, or with mathematics.'

Such warnings about failing to recognise the particular nature of designing are now common in design theory. Many people have especially warned against confusing design with science.

The scientific method is a pattern of problem-solving behaviour employed in finding out the nature of what exists, whereas the design method is a pattern of behaviour employed in inventing things of value which do not yet exist. Science is analytic; design is constructive. (Gregory, 1966)

The natural sciences are concerned with how things are... Design, on the other hand, is concerned with how things ought to be. (Simon, 1969)

To base design theory on inappropriate paradigms of logic and science is to make a bad mistake. Logic has interests in abstract forms. Science investigates extant forms. Design initiates novel forms. (March, 1976)

The emphasis in these admonitions is on the constructive, normative, creative nature of designing. Designing is a process of pattern synthesis, rather than pattern recognition. The solution is not simply lying there among the data, like the dog among the spots in the well known perceptual puzzle; it has to be actively constructed by the designer's own efforts.

Reflecting on his observations of urban designers, Levin (1966) commented that:

The designer knows (consciously or unconsciously) that some ingredient must be added to the information that he already has in order that he may arrive at an unique solution. This knowledge is in itself not enough in design problems, of course. He has to look for the extra ingredient, and he uses his powers of conjecture and original thought to do so. What then is this extra ingredient? In many if not most cases it is an 'ordering principle'. The preoccupation with geometrical patterns that is revealed in many town plans and many writings on the subject demonstrates this very clearly.

And of course it is not only in town planning, but in all fields of design, that one finds this preoccupation with geometrical patterns; a pattern (or some other ordering principle) seemingly *has* to be imposed in order to make a solution possible.

This pattern-constructing feature has been recognised as lying at the core of design activity by Alexander (1964, 1979), in his 'constructive diagrams' and 'pattern language'. The designer learns to think in this sketch-like form, in which the abstract patterns of user requirements are turned into the concrete patterns of an actual object. Hillier and Leaman (1976) suggested that it is like learning an artificial 'language', a kind of code which transforms 'thoughts' into 'words':

Those who have been trained as designers will be using just such a code ... which enables the designer to effect a translation from individual, organisational and social needs to physical artefacts. This code which has been learned is supposed to express and contain actual connections which exist between human needs and their artificial environment. In effect, the designer learns to 'speak' a language – to make a useful transaction between domains which are unlike each other (sounds and meanings in language, artefacts and needs in design) by means of a code or system of codes which structure that connection.

Designerly ways of knowing are embodied in these 'codes'. The details of the codes will vary from one design profession to another, but perhaps there is a 'deep structure' to design codes. We shall not know this until more effort has been made in externalising the codes.

What designers know about their own problem-solving processes remains largely tacit knowledge – *i.e.* they know it in the same way that a skilled person 'knows' how to perform that skill. They find it difficult to externalise their knowledge, and hence design education is forced to rely so heavily on an apprenticeship system of learning. It may be satisfactory, or at least understandable, for practicing designers to be inarticulate about their skills, but teachers of design have a responsibility to be as articulate as they possibly can about what it is they are trying to teach, or else they can have no basis for choosing the content and methods of their teaching.

Design Products

So far, I have concentrated on designerly ways of knowing that are embodied in the *processes* of designing. But there is an equally important area of knowledge embodied in the *products* of designing.

There is a great wealth of knowledge carried in the objects of our material culture. If you want to know how an object should be designed – *e.g.* what shapes and sizes it should have, what material it should be made from – go and look at existing examples of that kind of object, and simply copy (*i.e.* learn!) from the past. This, of course, was the 'design process' that was so successful in generating the material culture of craft society: the craftsperson simply copied the design of an object from its previous examples. Both Alexander (1964) and Jones (1970) have emphasised how the 'unselfconscious' processes of craft design led to extremely subtle, beautiful and appropriate objects. A very simple process can actually generate very complex products.

Objects are a form of knowledge about how to satisfy certain requirements, about how to perform certain tasks. And they are a form of knowledge that is available to everyone; one does not have to understand mechanics, nor metallurgy, nor the molecular structure of timber, to know that an axe offers (or 'explains') a very effective way of splitting wood. Of course, explicit knowledge about objects and about how they function *has* become available, and has sometimes led to significant improvements in the design of the objects. But in general, 'invention comes before theory' (Pye, 1978); the world of 'doing and making' is usually ahead of the world of understanding – technology leads to science, not *vice versa* as is often believed.

A significant branch of designerly ways of knowing, then, is the knowledge that resides in objects. Designers are immersed in this material culture, and draw upon it as the primary source of their thinking. Designers have the ability both to 'read' and 'write' in this culture: they understand what messages objects communicate, and they can create new objects which embody new messages. The importance of this two-way communication between people and 'the world of goods' has been recognised by Douglas and Isherwood (1979). In a passage that has strong connections to the arguments for a 'third area' of human knowledge in design, as distinct from the sciences and the humanities, they say:

For too long a narrow idea of human reasoning has prevailed which only accepts simple induction and deduction as worthy of the name of thinking. But there is a prior and pervasive kind of reasoning that scans a scene and sizes it up, packing into one instant's survey a process of matching, classifying and comparing. This is not to invoke a mysterious faculty of intuition or mental association. Metaphoric appreciation, as all the words we have used suggest, is a work of approximate measurement, scaling and comparison between like and unlike elements in a pattern.

'Metaphoric appreciation' is an apt name for what it is that designers are particularly skilled in, in 'reading' the world of goods, in translating back from concrete objects to abstract requirements, through their design codes. 'Forget that commodities are good for eating, clothing, and shelter,' Douglas and Isherwood say; 'forget their usefulness and try instead the idea that commodities are good for thinking; treat them as a nonverbal medium for the human relative faculty.'

Intrinsic Value of Design Education

The arguments for, and defence of, design in general education must rest on identifying the intrinsic values of design that make it justifiably a part of everyone's education. Above, I have tried to set out the field of 'designerly ways of knowing', as it relates to both the processes and the products of designing, in the hope that it will lead into an understanding of what these intrinsic values might be. Essentially, we can say that designerly ways of knowing rest on the manipulation of non-verbal codes in the material culture; these codes translate 'messages' either way between concrete objects and abstract requirements; they facilitate the constructive, solution-focused thinking of the designer, in the same way that other (*e.g.* verbal and numerical) codes facilitate analytic, problem-focused thinking; they are probably the most effective means of tackling the characteristically ill-defined problems of planning, designing and inventing new things.

From even a sketchy analysis, such as this, of designerly ways of knowing, we can indeed begin to identify features that can be justified in education as having intrinsic value. Firstly, we can say that design develops students' abilities in tackling a particular kind of problem. This kind of problem is characterised as ill-defined, or ill-structured, and is quite distinct from the kinds of well-structured problems that lie in the educational domains of the sciences and the humanities. We might even claim that our design problems are more 'real' than theirs, in that they are like the problems or issues or decisions that people are more usually faced with in everyday life.

There is therefore a strong educational justification for design as an introduction to, and assisting in the development of cognitive skills and abilities in real-world problem solving (Fox, 1981). We must be careful not to interpret this justification in instrumental terms, as a training in problem-solving skills, but in terms that satisfy the more rigorous criteria for education. As far as problem-solving is concerned, design in general education must be justified in terms of helping to develop an 'educated' person, able to understand the nature of ill-defined problems, how to tackle them, and how they differ from other kinds of problems. This kind of justification has been developed by McPeck (1981) in terms of the educational value of 'critical thinking'. A related justification is given by Harrison (1978), particularly in the context of practical design work, in terms of the radical connections between 'making and thinking'.

This leads us into a second area of justification for design in general education, based on the kind of thinking that is peculiar to design. This characteristically 'constructive' thinking is distinct from the more commonly acknowledged inductive and deductive kinds of reasoning. (March (1976) has related it to what the philosopher C. S. Peirce called 'abductive' reasoning.)

In educational terms, the development of constructive thinking must be seen as a neglected aspect of cognitive development in the individual. This neglect can be traced to the dominance of the cultures of the sciences and the humanities, and the dominance of the 'stage' theories of cognitive development. These theories, especially Piaget's, tend to suggest that the concrete, constructive, synthetic kinds of reasoning occur relatively early in child development, and that they are passed through to reach the higher states of abstract, analytical reasoning (*i.e.* the kinds of reasoning that predominate in the sciences, especially). There are other theories (for example, Bruner's) that suggest that cognitive development is a continuous process of interaction between different modes of cognition, all of which can be developed to high levels. That is, the qualitatively different types of cognition (*e.g.* 'concrete' and 'formal' types in Piaget's terms, 'iconic' and 'symbolic' in Bruner's terms) are not simply characteristic of different 'stages' of development, but are different kinds of innate human cognitive abilities, *all* of which can be developed from lower to higher levels.

The concrete/iconic modes of cognition are particularly relevant in design, whereas the formal/symbolic modes are more relevant in the sciences. If the 'continuous' rather than the 'stage' theories of cognitive development are adopted, it is clear that there is a strong justification for design education in that it provides opportunities particularly for the development of the concrete/iconic modes.

From this, we can move on to a third area of justification for design in general education, based on the recognition that there are large areas of human cognitive ability that have been systematically ignored in our educational system. Because most theorists of cognitive development are themselves thoroughly immersed in the scientific-academic cultures where numeracy and literacy prevail, they have overlooked the third culture of design. This culture relies not so much on verbal, numerical and literary modes of thinking and communicating, but on nonverbal modes. This is particularly evident in the designer's use of models and 'codes' that rely so heavily on graphic images – *i.e.* drawings, diagrams and sketches that are aids to internal thinking as well as aids to communicating ideas and instructions to others.

As well as these graphic models, there is also in design a significant use of mental imagery in 'the mind's eye' (Ferguson, 1977). The field of nonverbal thought and communication as it relates to design includes a wide range of elements, from 'graphicacy' to 'object languages', 'action languages' and 'cognitive mapping'. Most of these cognitive modes are strongest in the right hemisphere of the brain, rather than the left (Ornstein, 1975). So on this view the 'neglected area' of design in education is not merely one-third of human experience and ability, but nearer to one-half!

French (1979) has recognised nonverbal thinking as perhaps the principal justification for design in general education: 'It is in strengthening and uniting the entire nonverbal education of the child, and in its improvement of the range of acuity of his thinking, that the prime justification of the teaching of design in schools should be sought, not in preparing for a career or leisure, nor in training knowledgeable consumers, valuable as these aspects may be.'

The Discipline of Design

In this chapter I have taken up the argument put forward in the Royal College of Art report on 'Design in General Education' that there are 'designerly ways of knowing' that are at the core of the design area of education. First, I have stressed that we must seek to interpret this core of knowledge in terms of its intrinsic educational value, and not in the instrumental terms that are associated with traditional, vocational design education. Second, I have drawn upon the field of design research for what it has to say about the way designers work and think, and the kinds of problems they tackle. And third, I have tried to develop from this the justification that can be made for design as a part of general education in terms of intrinsic educational values.

I identified five aspects of designerly ways of knowing:

- Designers tackle 'ill-defined' problems.
- Their mode of problem-solving is 'solution-focused'.
- Their mode of thinking is 'constructive'.
- They use 'codes' that translate abstract requirements into concrete objects.
- They use these codes to both 'read' and 'write' in 'object languages'.

From these ways of knowing I drew three main areas of justification for design in general education:

- Design develops innate abilities in solving real-world, ill-defined problems.
- Design sustains cognitive development in the concrete/iconic modes of cognition.
- Design offers opportunities for development of a wide range of abilities in nonverbal thought and communication.

For me, something else also begins to emerge from these lines of argument. It seems to me that the design research movement and the design education movement are beginning to converge on what is, after all, their common concern – the discipline of design. The research path to design as a discipline has concentrated on understanding those general features of design activity that are common to all the design professions: it has been concerned with 'design in general' and it now allows us to generalise at least a little about the designerly ways of knowing. The education path to design as a discipline has also been concerned with 'design in general', and it has led us to consider what it is that can be generalised as of intrinsic value in learning to design. Both the research and the education paths, then, have been concerned with developing the general subject of design.

However, there is still a long way to go before we can begin to have much sense of having achieved a real understanding of design as a discipline – we have only begun to make rough maps of the territory. Following on from his comments on nonverbal education as the prime justification for design in general education, French (1979) also points out that there are certain implications arising from this:

> *If design teaching is to have this role it must meet certain requirements. It must 'stretch the mind', and ideally this involves a progression from step to step, some discipline of thought to be acquired in more or less specifiable components, reflected in a growing achievement of the pupil that both he and his teacher can recognise with some confidence. At present, there does not seem to be enough understanding, enough scholarly work on design, enough material of a suitable nature to make such teaching possible. I believe we should strive to remedy this state of affairs.*

The education path to design as a discipline forces us to consider the nature of this general subject of design, what it is that we are seeking to develop in the individual student, and how this development can be structured for learning. Like our colleagues in the sciences and the humanities we can at this point legitimately conclude that further research is needed! We need more research and enquiry: first into the designerly ways of knowing; second into the scope, limits and nature of innate cognitive abilities relevant to design; and third into the ways of enhancing and developing these abilities through education.

We need a 'research programme', in the sense in which Lakatos (1970) has described the research programmes of science. At its core is a 'touch-stone theory' or idea – in our case the view that 'there are designerly ways of knowing'. Around this core is built a 'defensive' network of related theories, ideas and knowledge – and I have tried to sketch in some of these in this chapter. In this way both design research and design education can develop a common approach to design as a discipline.

2

The Nature and Nurture of Design Ability

This chapter is in two parts. The first is concerned with the nature of design ability – the particular ways of thinking and behaving that designers, and all of us, adopt in tackling certain kinds of problems in certain kinds of ways. The second part is concerned with the nurture of design ability – that is, with the development of that ability through design education. My view is that through better understanding the nature of design ability, design educators may be better able to nurture it. I therefore see these two – nature and nurture – as complementary interests, and I do not intend to venture into those corners of psychology where fights go on over nature *versus* nurture in the context of general intelligence. However, I shall try to make a claim that design ability is, in fact, one of the several forms or fundamental aspects of human intelligence. It should, therefore, be an important element in everyone's education.

Nature

What Do Designers Do?

Everything we have around us – our environments, clothes, furniture, machines, communication systems, even much of our food – has been designed. The quality of that design effort therefore profoundly affects our quality of life. The ability of designers to produce efficient, effective, imaginative and stimulating designs is therefore important to all of us. And so it is important, first of all, to understand what it is that designers do when they exercise this ability.

Pragmatically, the most essential thing that any designer does is to provide, for those who will make a new artefact, a description of what that artefact should be like. Usually, little or nothing is left to the discretion of the makers – the designer specifies the artefact's dimensions, materials, finishes and colours. When a client

First presented as inaugural lecture as Professor of Design Studies, The Open University, 1989, and first published in *Design Studies* Vol 11, No 3, July 1990, pp. 127–140.

asks a designer for 'a design', that is what they want – the description. The focus of all design activity is that end-point.

The designer's aim, therefore, is the communication of a specific design proposal. Usually, this is in the form of a drawing or drawings, giving both an overview of the artefact and particular details. Even the most imaginative design proposals must usually be communicated in rather prosaic working drawings, lists of parts, and so on.

Sometimes, it is necessary to make full-scale mock-ups of design proposals in order that they can be communicated sufficiently accurately. In the motor industry, for example, full-scale models of new car bodies are made to communicate the complex three-dimensional shapes. These shapes are then digitized and the data communicated to computers for the production of drawings for making the body-panel moulds. Increasingly, in many industries, computerisation of both design and manufacture is substantially changing the mode of communication between designer and manufacturer, sometimes with the complete elimination of conventional detail drawings.

Before the final design proposal is communicated for manufacture, it will have gone through some form of testing, and alternative proposals may also have been tested and rejected. A major part of the designer's work is therefore concerned with the evaluation of design proposals. Again, full-scale models may be made – product manufacturing industries use them extensively for evaluating aesthetics, ergonomics, and consumer choice, as well as for production purposes. Small-scale 3D models are also often used in many industries – from architecture to chemical process plants.

However, drawings of various kinds are still the most extensively used modelling medium for evaluating designs – both informally in the designer's skilled reading of drawings and imagining their implications, and more formally in measuring dimensions, calculating stresses, and so on. In evaluating designs, a large body of scientific and technical knowledge can be brought to bear. This modelling, testing and modifying is the central, iterative activity of the design process.

Before a proposal can be tested, it has to be originated somehow. The generation of design proposals is therefore the fundamental activity of designers, and that for which they become famous or infamous. Although design is usually associated with novelty and originality, most run-of-the-mill designing is actually based on making variations on previous designs. Drawings again feature heavily in this generative phase of the design process, although at the earliest stages they will be just the designer's 'thinking with a pencil' and perhaps comprehensible only to him or her.

The kind of thinking that is going on is multi-facetted and multi-levelled. The designer is thinking of the whole range of design criteria and requirements set by the client's brief, of technical and legal issues, and of self-imposed criteria such as the aesthetic and formal attributes of the proposal. Often, the problem as set by the client's brief will be vague, and it is only by the designer suggesting possible solutions that the client's requirements and criteria become clear. The designer's very first conceptualizations and representations of problem and solution are therefore critical to the procedures that will follow – the alternatives that may be considered, the testing and evaluating, and the final design proposal.

Studies of Designing

Although there is such a great deal of design activity going on in the world, the nature of design ability is rather poorly understood. It has been taken to be a mysterious talent. However, for some years now, there has been a slowly growing body of understanding about the ways designers work and think, based on a wide variety of studies of designing (Cross, 1984). Some of these studies rely on the reports of designers themselves, such as those we have just seen, but there is also a broad spectrum running through observations of designers at work, experimental studies based on protocol analysis, to theorising about the nature of design ability.

Such studies often confirm the personal comments about practice made by designers themselves, but try also to add another layer of explanation of the nature of designing. For example, one feature of design activity that is frequently confirmed by such studies is the importance of the use of several initial, conjectured solutions by the designer. In his pioneering case studies of engineering design, Marples (1960) suggested that:

> *The nature of the problem can only be found by examining it through proposed solutions, and it seems likely that its examination through one, and only one, proposal gives a very biased view. It seems probable that at least two radically different solutions need to be attempted in order to get, through comparisons of sub-problems, a clear picture of the 'real nature' of the problem.*

This view emphasises the role of the conjectured solution as a way of gaining understanding of the design problem, and the need, therefore, to generate a variety of solutions precisely as a means of problem-analysis. It has been confirmed by Darke's (1979) interviews with architects, where she observed how they imposed a limited set of objectives or a specific solution concept as a 'primary generator' for an initial solution:

> *The greatest variety reduction or narrowing down of the range of solutions occurs early on in the design process, with a conjecture or concept-ualization of a possible solution. Further understanding of the problem is gained by testing this conjectured solution.*

The freedom – and necessity – of the designer to re-define the problem through the means of solution-conjecture was also observed in protocol studies of architects by Akin (1979), who commented:

> *One of the unique aspects of design behaviour is the constant generation of new task goals and redefinition of task constraints.*

It has been suggested that this feature of design behaviour arises from the nature of design problems: they are not the sort of problems or puzzles that provide all the necessary and sufficient information for their solution. Some of the relevant information can only be found by generating and testing solutions; some information, or 'missing ingredient', has to be provided by the designer himself, as noted by Levin (1966) from his observations of urban designers. Levin suggested that this extra ingredient is often an 'ordering principle' and hence we find the formal properties that are so often evident in designers' work, from towns designed as simple stars to teacups designed as regular cylinders.

However, designers do not always find it easy to generate a range of alternative solutions in order that they better understand the problem. Their 'ordering principles' or 'primary generators' can, of course, be found to be inappropriate, but designers often try to hang on to them, because of the difficulties of going back and starting afresh. From his case studies of architectural design, Rowe (1987) observed:

A dominant influence is exerted by initial design ideas on subsequent problem-solving directions... Even when severe problems are encountered, a considerable effort is made to make the initial idea work, rather than to stand back and adopt a fresh point of departure.

This tenacity is understandable but undesirable, given the necessity of using alternative solutions as a means of understanding the 'real nature' of the problem. However, Waldron and Waldron (1988), from their engineering design case study, came to a more optimistic view about the 'self-correcting' nature of the design process:

The premises that were used in initial concept generation often proved, on subsequent investigation, to be wholly or partly fallacious. Nevertheless, they provided a necessary starting point. The process can be viewed as inherently self-correcting, since later work tends to clarify and correct earlier work.

It becomes clear from these studies of designing that architects, engineers, and other designers adopt a problem-solving strategy based on generating and testing potential solutions. In a laboratory experiment based on a specific problem-solving task, Lawson (1979) compared the strategies of architects with those of scientists, and found a noticeable difference, in that '[The scientists] operated what might be called a problem-focussing strategy... architects by contrast adopted a solution-focussing strategy.'

In a supplementary experiment, Lawson found that these different strategies developed during the architects' and scientists' education; whilst the difference was clear between postgraduate students, it was not clear between first-year undergraduate students. The architects had therefore learned their solution-focussing strategy, during their design education, as an appropriate response to the problems they were set. This is presumably because design problems are inherently ill-defined, and trying to define or comprehensively to understand the problem (the scientists' approach) is quite likely to be fruitless in terms of generating an appropriate solution within a limited timescale.

The difference between a scientific approach and a design approach to problem solving has also been emphasised in theoretical studies, such as that of March (1976), who pointed out that:

Logic has interests in abstract forms. Science investigates extant forms. Design initiates novel forms. A scientific hypothesis is not the same thing as a design hypothesis. A logical proposition is not to be mistaken for a design proposal. A speculative design cannot be determined logically, because the mode of reasoning involved is essentially abductive.

This 'abductive' reasoning is a concept from the philosopher Peirce, who distinguished it from the other more well-known modes of inductive and deductive reasoning. Peirce (quoted by March) suggested that 'Deduction proves that something must be; induction shows that something actually is operative; abduction merely suggests that something may be.' It is therefore the logic of conjecture. March prefers to use the term 'productive' reasoning. Others, such as Bogen (1969), have used terms such as 'appositional' reasoning in contra-distinction to propositional reasoning.

Design ability is therefore founded on the resolution of ill-defined problems by adopting a solution-focussing strategy and productive or appositional styles of thinking. However, the design approach is not necessarily limited to ill-defined problems. Thomas and Carroll (1979) conducted a number of experiments and protocol studies of designing and concluded that a fundamental aspect is the nature of the approach taken to problems, rather than the nature of the problems themselves:

Design is a type of problem solving in which the problem solver views the problem or acts as though there is some ill-definedness in the goals, initial conditions or allowable transformations.

There is also, of course, the reliance in design upon the media of sketching, drawing and modelling as aids to the generation of solutions and to the very processes of thinking about the problem and its solution. The process involves what Schön (1983) has called 'a reflective conversation with the situation'. From his observations of the way design tutors work, Schön commented that, through sketches,

[The designer] shapes the situation, in accordance with his initial appreciation of it; the situation 'talks back', and he responds to the back-talk.

Design ability therefore relies fundamentally on non-verbal media of thought and communication. There may even be distinct limits to the amount of verbalising that we can productively engage in about design ability. Daley (1982) has suggested that:

The way designers work may be inexplicable, not for some romantic or mystical reason, but simply because these processes lie outside the bounds of verbal discourse: they are literally indescribable in linguistic terms.

For design researchers this is a worrying conclusion. However, this brief review of studies of designing does enable us at least to summarise the core features of design ability as comprising abilities to:

- resolve ill-defined problems
- adopt solution-focussing strategies
- employ abductive/productive/appositional thinking
- use non-verbal, graphic/spatial modelling media.

Design Ability is Possessed by Everyone

Although professional designers might naturally be expected to have highly developed design abilities, it is also clear that non-designers also possess at least some aspects, or lower levels of design ability. Everyone makes decisions about arrangements and combinations of clothes, furniture, *etc.* – although in industrial societies it is rare for this to extend beyond making selections from available goods that have already been designed by someone else.

However, in other societies, especially non-industrial ones, there is often no clear distinction between professional and amateur design abilities – the role of the professional designer may not exist. In craft-based societies, for example, craftspeople make objects that are not only highly practical but often also very beautiful. They would therefore seem to possess high levels of design ability – although in such cases, the ability is collective rather than individual: the beautiful-functional objects have evolved by gradual development over a very long time, and the forms of the objects are rigidly adhered to from one generation to the next.

Even in industrial societies, with a developed class of professional designers, there are often examples of vernacular design persisting, usually following implicit rules of how things should be done, similar to craftwork. Occasionally there are examples of 'naive' design breaking out in industrial societies, with many of the positive attributes that 'naive' art has. A classic example is the 'Watts Towers' – an environmental fantasy created by Simon Rodia in his Los Angeles backyard between the nineteen-twenties and -fifties. In architecture and planning, there have been moves to incorporate non-professionals into the design process – through design participation or community architecture. Although the experiments have not always been successful – in either process or product – there is at least a recognition that the professionals could, and should, collaborate with the non-professionals. Knowledge about design is certainly not exclusive to the professionals.

A strong indication of how widespread design ability is comes from the introduction of design as a subject in schools. It is clear from the often very competent design work of schoolchildren of all ages that design ability is inherent in everyone.

Design Ability Can Be Damaged or Lost

Although some aspects of design ability can be seen to be widespread in the general population, it has also become clear that the cognitive functions upon which design ability depends can be damaged or lost. This has been learned from experiments and observations in the field of neuropsychology, particularly the work which became known as 'split-brain' studies, described by Gazzaniga (1970).

These studies showed that the two hemispheres of the brain have preferences and specialisations for different types of perceptions and knowledge. Normally, the large bundle of nerves (the corpus callosum) which connects the two hemispheres ensures rapid and comprehensive communication between them, so that it is impossible to study the workings of either hemisphere in isolation from its mate. However, in order to cure epilepsy, some people have had their corpus callosum surgically severed, and became subjects for some remarkable experiments to investigate the isolated functions of the two hemispheres (Sperry et al., 1969).

Studies of other people who had suffered damage to one or other hemisphere had already revealed some knowledge of the different specialisations. In the main, these studies had shown the fundamental importance of the left hemisphere – it controlled speech functions and the verbal reasoning normally associated with logical thought. The right hemisphere appeared to have no such important functions. Indeed, the right became known as the 'minor' hemisphere, and the left as the 'major' hemisphere. Nevertheless, there is an equal sharing of control of the body; the left hemisphere controls the right side, and vice versa, for some perverse reason known only to the Grand Designer in the Sky.

This left-right crossover means that sensory reception on the left side of the body is communicated to the brain's right hemisphere, and vice versa. This even applies, in a more complex way, to visual reception; it is not simply that the left eye communicates with the right hemisphere, and vice versa, but that, for both eyes, reception from the left visual field is communicated to the right hemisphere, and vice versa. Ingenious experiments were therefore devised in which visual stimuli could be sent exclusively to either the left or right hemisphere of the split-brain subjects.

These experiments showed that the separated hemispheres could receive, and therefore 'know', separate items of information. The problem was how to get the hemispheres to communicate what they knew back to the experimenter. The left hemisphere, of course, can communicate verbally, but the right hemisphere is mute. Some experimenters resolved this problem by visually communicating a word or image to the right hemisphere, and asking it to identify a matching object by touch with the left hand.

From experiments such as these, neuropsychologists such as Blakeslee (1980) developed a much better understanding of the functions and abilities of the right hemisphere. Although mute, it is by no means stupid, and it perceives and knows things that the left hemisphere does not. In general, this is the kind of knowledge that we categorise as intuitive. The right hemisphere excels in emotional and aesthetic perception, in the recognition of faces and objects, and in visuo-spatial and constructional tasks. This scientific, rational evidence therefore supports our own personal, intuitive understanding of ourselves, and also supports the (often poorly articulated) view of artists and many designers that verbalisation (*i.e.* allowing the left hemisphere to dominate) obstructs intuitive creation.

It is now known that damage to the right hemisphere can impair brain functions that relate strongly to intuitive, artistic and design abilities. This has been confirmed by studies of, for instance, drawing ability. One classic case is that of an artist who suffered right-brain damage. Although he could make an adequate sketch of an object such as a telephone when he had it in front of him, he could not draw the same object from memory and resorted instead to 'reasoning' about what such an object might be like – producing strange new 'designs'. Studies of split-brain subjects have also shown, in general, that they can draw better with their left hand (even though they are not naturally left-handed people) than their right. Recognition of this right-brain ability has been put to constructive use in art education by Edwards (1979), who trains students to 'draw on the right side of the brain'. Anita Cross (1984) has drawn attention to the relevance of the 'split-brain' studies to improving our understanding of design ability.

There is, of course, a long history of studies in psychology of cognitive styles, which are usually polarised into dichotomies such as

- convergent – divergent
- focussed – flexible
- linear – lateral
- serialist – holist
- propositional – appositional.

Such natural dichotomies may reflect the underlying dual structure of the human brain and its apparent dual modes of information processing. Cross and Nathenson (1981) have drawn attention to the importance of understanding cognitive styles for design education and design methodology.

Design as a Form of Intelligence

What I have attempted to show is that design ability is a multi-faceted cognitive skill, possessed in some degree by everyone. I believe that there is enough evidence to make a reasonable claim that there are particular, 'designerly' ways of knowing, thinking and acting. In fact, it seems possible to make a reasonable claim that design ability is a form of natural intelligence, of the kind that the psychologist Howard Gardner (1983) has identified. Gardner's view is that there is not just one form of intelligence, but several, relatively autonomous human intellectual competences. He distinguishes six forms of intelligence:

- linguistic
- logical-mathematical
- spatial
- musical
- bodily-kinaesthetic
- personal.

Aspects of design ability seem to be spread through these six forms in a way that does not always seem entirely satisfactory. For example, spatial abilities in problem-solving (including thinking 'in the mind's eye') are classified by Gardner under spatial intelligence, whereas many other aspects of practical problem-solving ability (including examples from engineering) are classified under bodily-kinaesthetic intelligence. In this classification, the inventor appears alongside the dancer and the actor, which doesn't seem appropriate. It seems reasonable, therefore, to try to separate out design ability as a form of intelligence in its own right.

Gardner proposes a set of criteria against which claims for a distinct form of intelligence can be judged. These criteria are as follows, with my attempts to match 'design intelligence' against them.

• *Potential isolation by brain damage.* Gardner seeks to base forms of intelligence in discrete brain-centres, which means that particular faculties can be destroyed (or spared) in isolation by brain damage. The evidence here for design intelligence draws upon the work with 'split-brain' and brain-damaged patients, which shows that abilities such as geometric reasoning, 3-dimensional problem solving and visuo-spatial thinking are indeed located in specific brain-centres.

• *The existence of* idiots savants*, prodigies and other exceptional individuals.* Here, Gardner is looking for evidence of unique abilities which sometimes stand out in individuals against a background of retarded or immature general intellectual development. In design, there are indeed examples of otherwise ordinary individuals who demonstrate high levels of ability in forming their own environments – the 'naive' designers.

• *An identifiable core operation or set of operations.* By this, Gardner means some basic mental information-processing operation(s) which deal with specific kinds of input. In design, this might be the operation of transforming the input of the problem brief into the output of conjectured solutions, or the ability to generate alternative solutions. Gardner suggests that 'Simulation on a computer is one promising way of establishing that a core operation exists.' Work in artificial intelligence on the generation of designs by computer is therefore helping to clarify the concept of a natural design intelligence.

• *A distinctive developmental history, and a definable set of expert, end-state performances.* This means recognisable levels of development or expertise in the individual. Clearly, there are recognisable differences between novices and experts in design, and stages of development amongst design students. But a clarification of the developmental stages of design ability is something that we still await, and is sorely needed in design education.

• *An evolutionary history.* Gardner argues that the forms of intelligence must have arisen through evolutionary antecedents, including capacities that are shared with other organisms besides human beings. In design, we do have examples of animals and insects that construct shelters and environments, and use and devise tools. We also have the long tradition of vernacular and craft design as a precursor to modern, innovative design ability.

• *Susceptibility to encoding in a symbol system.* This criterion looks for a coherent, culturally-shared system of symbols which capture and communicate information relevant to the form of intelligence. Clearly, in design we have the use of sketches, drawings and other models which constitute a coherent, symbolic media system for thinking and communicating.

- *Support from experimental psychological tasks.* Finally, Gardner looks for evidence of abilities that transfer across different contexts, of specific forms of memory, attention or perception. We only have a few psychological studies of design behaviour or thinking, but aspects such as solution-focused thinking have been identified. More work in this area needs to be done.

If asked to judge the case for design intelligence on this set of criteria, we might have to conclude that the case is 'not proven'. Whilst there is good evidence to meet most of the criteria, on some there is a lack of substantial or reliable evidence. However, I think that viewing designing as a 'form of intelligence' is productive; it helps to identify and clarify features of the nature of design ability, and it offers a framework for understanding and developing the nurture of design ability.

Nurture

Learning to Design

How do people learn to design, and on what principles should design education be based? Clearly, some development of design ability does take place in students – certainly at the level of tertiary, professional education, where we can compare the work of the same student over the years of his or her course. The crude, simple work of the first-year student develops into sophisticated, complex work by the final year. But the educational processes which nurture this development are poorly understood – if at all – and rely heavily on the project method.

In pre-industrial society, there was really no such thing as design education. People learned to make products in learning the skills of a trade, they were apprenticed to a master craftsperson, and they learned to copy. In many respects, the old tradition of design education, derived from the Beaux Arts School, was based on apprenticeship. Students worked closely with a master; they learned set responses to set problems; products and processes were predictable.

Modern, industrial design education owes much to the experimental work of the Bauhaus – the German design school of the nineteen-twenties and -thirties – in particular, the radical 'basic course' introduced by Johannes Itten. As Anita Cross (1983) has suggested, many of the basic course's educational principles may well have been developed from, or influenced by the work of educational innovators such as Froebel, Montessori and Dewey. The Bauhaus also integrated design education with aesthetic cultures such as dance, theatre and music, as well as cultures of technology and industry. Itten himself incorporated physical exercises into his courses, and required his students, for example, to swing their arms and bodies in circular movements before attempting to draw freehand circles. He and other tutors also encouraged tactile perception and the construction of collages from randomly-collected junk and other materials. From what we now know of the development of the thought-modes of the right hemisphere of the brain, these nonverbal, tactile, analogical experiences were intuitively correct aspects of design education.

Most of the Bauhaus innovations are now severely watered-down in conventional design education, usually retaining just a few vestiges of exercises in colour, form and composition. With the possible exception of the Hochschule für Gestaltung (HfG) at Ulm in the nineteen-sixties, there have been no comparable innovations in curriculum development in design education since the Nazis closed the Bauhaus in 1933.

In general education it is particularly important that teachers have a fundamental understanding of the abilities that they are seeking to develop in their students. In tertiary, professional education, teachers can get by as long as their students are reasonably competent enough to enter their profession at the end of their course. In professional education the distinctions between education and training are perhaps less clear-cut than they are in general education, where no particular profession is the goal. Professional education has instrumental, or extrinsic aims, whereas general education has to pursue intrinsic aims that are somehow inherently good for the individual.

I suggest that it is through understanding the nature of design ability that we can begin to construct an understanding of the intrinsic values of design education. For example, we can make a strong justification for design based on its development of personal abilities in resolving ill-defined problems – which are quite different from the well-defined problems dealt with in other areas of the curriculum. We can also justify the designer's solution-focused strategies and appositional thinking styles as promoting a certain type of cognitive development – in educational terms, the concrete/iconic modes that are often assumed to be the 'earlier' or 'minor' modes of cognition, and less important than the formal/symbolic modes. Furthermore, there is a sound justification in the educational value of design in its development of the whole area of non-verbal thought and communication.

Design Education in the Open

To attempt design education 'at a distance' has been a great challenge for us at The Open University. At first, we had real doubts about how to teach design through the new distance-learning systems, including teaching-texts, TV and radio, and more recently computers and information technologies. We were not alone: there were certainly those who said that it could not be done, given the reliance on face-to-face or over-the-drawing-board teaching in conventional design education. However, as the design work of many of our OU students now shows, some development of design ability does take place through our distance-learning courses, just as it does in conventional courses. It also seems increasingly apparent that an *open* version of design education can offer a new, universal model more appropriate to our emerging post-industrial society and technology.

I suggest that there are four key aspects or foundations to develop for a universal, open version of design education: making design education accessible, ubiquitous, continuous and explicit.

Making design education *accessible* means making it available to everyone. In many countries, design is now a part of general education – it is taught in schools to children. This means that design education is no longer just a preparation for a profession, but is recognised as having intrinsic value in the development of everyone's intellect. It has become a part of our individual and collective intellectual culture, just like literature, science or mathematics; it has become a part of basic educational proficiency, just like reading, writing and numeracy.

Making design education *ubiquitous* means making it available everywhere. As with many other aspects of society, culture and technology, a ubiquitous design education is being facilitated by information technology, computers and the internet. It is no longer necessary to be physically present in a design studio – neither in professional practice nor in education. Virtual studios and virtual universities can be open to everyone, around the world and round the clock.

And just as education no longer stops at a certain time of day, it no longer stops at a certain age; accessibility and ubiquity also mean that education must be *continuous* and available throughout one's lifetime. Education is now recognised as a life-long process, something that each of us can engage in at any age, whether from personal choice or the necessity of keeping up to date, well informed and well educated in the changing skills and knowledge of post-industrial technology.

For these reasons, if for no other, it becomes necessary to make design education more *explicit*. Making design education explicit calls for a new kind of pedagogical approach from design teachers. The increased attention on design education in recent years has exposed the lack of clearly articulated and well understood principles of design education.

In developing a post-industrial view of design education it will be particularly important that teachers have a fundamental understanding of the underlying, intrinsic abilities that they are seeking to develop in their students. We need a secure foundation from which to question the relevance of conventional skills. We have moved beyond the apprenticeship system of pre-industrial design, and we must move beyond the pupilage system of industrial design education. We need to base design education on tested theories from education, psychology and cognitive science, and from design research, and we need a much stronger experimental base for educational innovation.

The Development of Design Ability

Although it may be present to some degree in everyone, design ability seems stronger in some people than others, and also seems to develop with experience. Experienced designers are able to draw on their knowledge of previous exemplars in their field of design, and they also seem to have learned the value of rapid problem-exploration through solution-conjecture. They use early solution attempts as experiments to help identify relevant information about the problem. In comparison, novice designers often become bogged-down in attempts to understand the problem before they start generating solutions.

Another difference between novices and experts is that novices will often pursue a 'depth-first' approach to a problem – sequentially identifying and exploring sub-solutions in depth, and amassing a number of partial sub-solutions that then somehow have to be amalgamated and reconciled, in a 'bottom-up' process. They can also become 'fixated' on a particular solution possibility. Experts usually pursue predominantly 'breadth-first' and top-down strategies, and are more willing to reject an early solution when it is discovered to be fundamentally flawed.

Conventional wisdom about the nature of expertise in problem-solving seems often to be contradicted by the behaviour of expert designers. But designing has many differences from conventional problem-solving, in which there is usually a single, correct solution to the problem. In design education we must therefore be very wary about importing models of behaviour from other fields. Empirical studies of design activity have frequently found 'intuitive' features of design ability to be the most effective and relevant to the intrinsic nature of design. Some aspects of design theory, however, have tried to develop counter-intuitive models and prescriptions for design behaviour. We still need a much better understanding of what constitutes expertise in design, and how we might assist novice students to gain that expertise.

In contrast to the artistic, intuitive procedures encouraged by the Bauhaus, design education has more recently concentrated on teaching more rational, systematic approaches. Some aspects of design ability have been codified into 'design methods' (Cross, 1989). Without those methods, it would have been much harder for us at The Open University to clarify and to try to teach some elements of design ability. Because skilled designers in practice often appear to proceed in a rather *ad-hoc* and unsystematic way, some people claim that learning a systematic process does not actually help student designers. However, a study by Radcliffe and Lee (1989) did show that a systematic approach can be helpful to students. They found that the use of more 'efficient' design processes (following closer to an 'ideal' sequence) correlated positively with both the quantity and the quality of the students' design results. Other studies have tended to confirm this.

Designing is a form of skilled behaviour. Developing any skill usually relies on controlled practice and the development of technique. The performance of a skilled practitioner appears to flow seamlessly, adapting the performance to the circumstances without faltering. But learning is not the same as performing, and underneath skilled performance lies mastery of technique and procedure. The design student needs to develop a strategic approach to the overall process, based on some simple but effective techniques or methods. What I hope we shall achieve through continued studies of the nature and nurture of design ability is that design education will become a reliably successful means for the development of design ability in everyone.

3

Natural and Artificial Intelligence in Design

A common joke is to say that 'the opposite of artificial intelligence is natural stupidity'. In the frustrations of our everyday life, most of us think that designers such as architects, product designers and computer software designers do indeed display considerable evidence of natural stupidity in the results of their work. But in this chapter I want to relate the subject of artificial intelligence (AI) in design, not to the natural stupidity, but to the natural *intelligence* in design that is possessed certainly by good designers, and in fact is possessed to some extent by all of us. My starting point is that people are designers – and some people are very good designers.

Designing is something that all people do; something that distinguishes us from other animals, and (so far) from machines. The ability to design is a part of human intelligence, and that ability is natural and widespread amongst the human population. We human beings have a long history of design ability, as evidenced in the artefacts of previous civilisations and in the continuing traditions of vernacular design and traditional craftwork. The evidence from different cultures around the World, and from designs created by children as well as by adults, suggests that everyone is capable of designing.

But we also know that some people are better designers than others. Ever since the emergence of designers as separate professions, it appears that some people have a design ability that is more highly developed than other people – either through some genetic endowment or through social and educational development. In fact, some people are very good at designing.

But can a machine design? That of course is the question that concerns researchers in the artificial intelligence of design. It is a question I shall return to, but first let me try to present something about what has been learned about the natural intelligence of design, and especially about some of those people who are very good at designing.

First presented as the keynote speech at *AI in Design*, Lisbon, Portugal, 1998, and first published as 'Natural Intelligence in Design', *Design Studies*, Vol 20, No 1, January 1999, pp. 25–39.

Research in Design Thinking

For many years now there has been a rather embarrassingly slow but nonetheless steady growth in our understanding of design ability The pioneer research paper in this field was the study of engineering designers by Marples (1960). A decade later, we had Eastman's (1970) also pioneering protocol studies of architects, and it was in the 1970s that we saw the first significant growth of the new field of design research. I collected some of the early examples of design research together in a book on *Developments in Design Methodology* (Cross, 1984), and more recently co-edited another review on *Research in Design Thinking* (Cross *et al.,* 1992).

The kinds of methods for researching the nature of design thinking that have been used have included:

- *Interviews with designers*
These have usually been with designers who are acknowledged as having well-developed design ability, and the methods have usually been unstructured interviews which sought to obtain these designers' reflections on the processes and procedures they use – either in general, or with reference to particular works of design. Examples include Lawson (1994) and Cross and Clayburn Cross (1996).

- *Observations and case studies*
These have usually been focused on one particular design project at a time, with observers recording the progress and development of the project either contemporaneously or post-hoc. Both participant and non-participant observation methods have been included, and varieties of real, artificially-constructed and even re-constructed design projects have been studied. Examples include Candy and Edmonds (1996), Galle (1996) and Valkenburg and Dorst (1998).

- *Protocol studies*
This more formal method has usually been applied to artificial projects, because of the stringent requirements of recording the protocols – the 'thinking-aloud' and associated actions of subjects asked to perform a set design task. Both inexperienced (often student) designers and experienced designers have been studied in this way. Examples include Lloyd and Scott (1994), Gero and McNeill (1998), and the Delft Protocols Workshop (Cross, *et al.*, 1996).

- *Reflection and theorising*
As well as the empirical research methods listed above, there has been a significant history in design research of theoretical analysis and reflection upon the nature of design ability. Leading examples are Simon (1969) and Schön (1983).

- *Simulation trials*
A relatively new development in research methodology has been the attempt of AI researchers to simulate human thinking through artificial intelligence techniques. Although AI techniques may be meant to supplant human thinking, research in AI can also be a means of trying to understand human thinking. Many

examples have been included in the proceedings of the series of *AI in Design* conferences, starting in 1991.

We therefore have a varied set of methods which have been used for research into design thinking. The set ranges from the more abstract to the more concrete types of investigation, and from the more close to the more distant study of actual design practice. The studies have ranged through inexperienced or student designers, to experienced and expert designers, and even on to forms of non-human, artificial intelligence. All of these methods have helped researchers to develop insights into what I am referring to as 'designerly' ways of thinking, or the natural intelligence of design.

Personally, I am particularly interested in what the best, expert designers have to say about design, because they may help us to develop insights into what it means to think, not just like any of us, but like a *good* designer. Therefore I am going to use some quotations from expert, outstanding designers, to illustrate the kinds of things that they say when they are interviewed about design thinking. The examples come from different design domains – architecture, engineering and product design – and I am going to relate what they say to the insights about the nature of good design thinking that the design researchers have begun to compile in recent years.

What Expert Designers Say About Designing

A famous example of early Modern Architecture was the 1930 'Tugendhat House' in Brno, Czechoslovakia, designed by Ludwig Mies van der Rohe. Apparently, according to Mies, the client had approached the architect after seeing some of the rather more conventional houses that he had designed before. Then, Mies said, in an anecdote reported by Simon (1969), when he showed the surprising new design to the client,

> *He wasn't very happy at first. But then we smoked some good cigars, ... and we drank some glasses of a good Rhein wine, ... and then he began to like it very much.*

The difficult lesson that we have to learn from this example, I believe, is that *design is rhetorical.* By this, I mean that design is persuasive. You, like me, have probably experienced this for yourself – for example, you go to a car sale, looking for a sensible, modest car, and come away with something that is impractical but beautiful! Perhaps the most famous example worldwide was the design of the Sony Walkman – a product that none of us realised we wanted, until we saw it.

Design is rhetorical also in the sense that the designer, in constructing a design proposal, constructs a particular kind of argument, in which a final conclusion is developed and evaluated as it develops against both known goals and previously unsuspected implications. This rhetorical nature of design has been summarised in a comment by the outstanding architect, Denys Lasdun (1965):

Our job is to give the client ... not what he wants, but what he never dreamed he wanted; and when he gets it, he recognizes it as something he wanted all the time.

I think that we should try to see through the apparent arrogance in this statement, to the underlying truth that clients do want designers to transcend the obvious and the mundane, and to produce proposals which are exciting and stimulating as well as merely practical. What this means is that design is not a search for the optimum solution to the given problem, but that *design is exploratory.* The creative designer interprets the design brief not as a specification for a solution, but as a kind of partial map of unknown territory (as suggested by Jones, 1970), and the designer sets off to explore, to discover something new, rather than to return with yet another example of the already familiar.

The vagueness, or slipperiness of the relationship between problem and solution in designing is also conveyed in the comment of the furniture designer Geoffrey Harcourt, discussing how one of his particular designs emerged:

As a matter of fact, the solution that I came up with wasn't a solution to the problem at all ... But when the chair was actually put together, in a way it solved the problem quite well, but from a completely different point of view. (Quoted by Davies, 1985.)

His comment suggests something of the perceptual aspect of design thinking – seeing the vase rather than the faces, in the well-known ambiguous figure. It implies that *design is emergent* – relevant features emerge in putative solution concepts, and can be recognised as having properties that suggest how the developing solution-concept might be matched to the developing problem-concept. In design, the solution and the problem develop together.

The ill-defined nature of design problems means that they cannot be solved simply by collecting and synthesising information, as the architect Richard MacCormac (1976) has observed:

I don't think you can design anything just by absorbing information and then hoping to synthesise it into a solution. What you need to know about the problem only becomes apparent as you're trying to solve it.

MacCormac is saying that all the relevant information cannot be predicted and established in advance of the design activity. The directions that are taken during the exploration of the design territory are influenced by what is learned along the way, and by the partial glimpses of what might lie ahead. In other words, *design is opportunistic*, and so the path of exploration cannot be predicted in advance.

Given the apparently ad hoc and surprise-full nature of creative design activity, it is not unusual for designers, when talking about design thinking, to refer to the role of 'intuition' in their reasoning processes. For instance, the industrial designer Jack Howe has commented:

I believe in intuition. I think that's the difference between a designer and an engineer... I make a distinction between engineers and engineering designers... An engineering designer is just as creative as any other sort of designer. (Quoted by Davies, 1985.)

This emphasis on 'intuition' is perhaps a bit surprising, coming from someone with a reputation for rather severe, rational design work. But I think that the concept of 'intuition' is a convenient, shorthand word for what really happens in design thinking. The more useful concept that has been used by design researchers in explaining the reasoning processes of designers is that *design is abductive:* a type of reasoning different from the more familiar concepts of inductive and deductive reasoning, but which is the necessary logic of design – the necessary but difficult step from function to form (Roozenburg, 1993).

The thinking processes of the designer seem to hinge around the relationship between internal mental processes and their external expression and representation in sketches. As the engineer-architect Santiago Calatrava has said:

To start with you see the thing in your mind and it doesn't exist on paper and then you start making simple sketches and organising things and then you start doing layer after layer ... it is very much a dialogue. (Quoted by Lawson, 1994.)

Acknowledging the dialogue or 'conversation' that goes on between internal and external representations is part of the recognition that *design is reflective.* The designer has to have some medium – which is the sketch – which enables half-formed ideas to be expressed and to be reflected upon: to be considered, revised, developed, rejected and returned to.

Given the complex nature of design thinking, therefore, it hardly seems surprising that the structural engineering designer Ted Happold should have suggested that:

I really have, perhaps, one real talent; which is that I don't mind at all living in the area of total uncertainty. (Quoted by Davies, 1985.)

Happold certainly needed this talent, as a leading member of the structural design team for some of the most challenging buildings in the world, such as the Sydney Opera House and the Pompidou Centre in Paris. The uncertainty of design is both the frustration and the joy that designers get from their activity; they have learned to live with the fact that *design is ambiguous.* Designers will generate early tentative solutions, but also leave many options open for as long as possible; they are prepared to regard solution concepts as necessary, but imprecise and often inconclusive.

One final theme that emerges from conversations with designers is the sense of risk-taking that accompanies creative design. The racing car designer Gordon Murray has said:

There are patches of quite – loneliness, really, when you sit there and you think – I'm committed to this crazy idea! (Quoted by Cross and Clayburn Cross, 1996.)

This comment is from someone whose innovative, Formula One cars for the Brabham and McLaren teams won more than 50 Grand Prix races, and four Drivers' World Championships, and who designed the outstanding McLaren F1 – a road-going 'supercar' that also went on to dominate in GT racing for several seasons. In the face of the uncertainty of original design, there comes a time when the designer has to make a personal commitment. I think we have to acknowledge that *design is risky* – it is not comfortable, and it is not easy. The designers I have quoted above – all of them successful, and acknowledged expert or even outstanding designers – made their reputations by taking risks.

In quoting these designers, and interpreting those quotations in terms of concepts from design research, I have been trying to show two things. Firstly, that although designers themselves do not normally use the kinds of concepts that the researchers use, we are talking about the same experiences and perceptions: we are talking about – hopefully, developing a disciplined conversation about (see Cross, 1999) – the natural intelligence of highly-developed design ability. Secondly, I have wanted to show that this is a difficult conversation: we are not talking about simple activities that can be expressed in simple concepts.

I don't want to imply that designing is mysterious and obscure; but I do want to show that it is complex. Although everyone can design, designing is one of the highest forms of human intelligence.

The Role of Sketching in Design

In order to explore the complexity of design thinking a little more deeply, I would like to consider how and why designers use one of the particular tools that helps them to think – the tool of sketching. The use of sketches is clearly an important part of the natural processes of designing, but trying to understand just what this importance is, is something that has only relatively recently become a subject of more careful consideration and analysis by design researchers.

I have already used a quotation from Santiago Calatrava to the effect that sketching is fundamental, as a kind of 'dialogue' situation for the designer. But why is it necessary for designers to draw at all? One obvious reason is that the end point of the design process usually requires a drawing, or a set of drawings, that provide a model of the object – the building or the product – that is to be made by the builder or manufacturer. That is the designer's goal – to provide that model. If, given the brief for a new product, the designer could immediately make that final model, then there would really be no need for a design process at all – the designer would simply read the brief and then prepare the final drawings.

Perhaps that is the goal of some the research in AI in design: to construct a machine that takes the design brief as input and gives the design drawings (or probably some other form of data for manufacture) as output. But human beings do not seem to be able to do that. A design process is necessary in which the final drawings are gradually, and sometimes rather arduously approached through a series of other drawings that we call sketches. As the industrial designer Jack Howe said, about how to start a design project, or how to proceed when stuck, 'I draw *something* – even if it's "potty" – the act of drawing seems to clarify my thoughts' (quoted by Davies, 1985). Trying to understand what goes on in that sketching-and-thinking process should give us some insights into the nature of the design process.

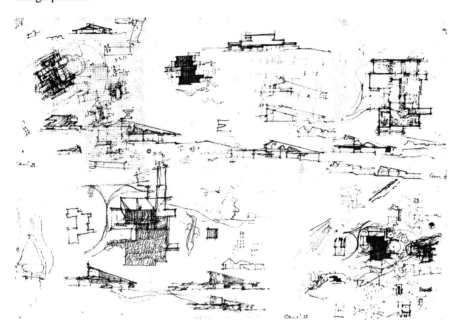

Figure 3.1. Alvar Aalto: design sketches for the Maison Carré, Bazoches, France

Drawing and sketching have been used in design for a long time – certainly since long before the Renaissance, although it is the period since that time that has seen a massive growth in the use of drawings, as designed objects have become more complex and more novel. Many of Leonardo da Vinci's drawings of machines show one of the key aspects of drawings, in terms of their purpose of communicating to someone else how a new product should be built, and also how it should work. Some of Leonardo's design drawings also show how a drawing can be not only a communication aid, but also a thinking and reasoning aid. Tzonis (1992) has discussed how Leonardo's sketches for the design of fortifications show how he used sight-lines and missile trajectories as lines to set up the design of the fortifications, and how cognitive processes were assisted by drawing.

Similar kinds of drawing and thinking can also be seen in more contemporary sketches, such as those of the architect Alvar Aalto (Figure 3.1). Apparently Aalto would sometimes use random drawing marks as stimuli to the development of ideas for building forms. But in this set of design sketches we see how sketching can help the designer to consider many aspects at once – we see plans, elevations, sections, details, all being drawn together and thus all being thought about, reasoned about, all together, alongside calculations of areas, volumes, and perhaps costs.

The architect Richard MacCormac has said, 'I use drawing as a process of criticism and discovery' (quoted by Lawson, 1994). The concepts that are drafted in design sketches are there to be criticised, not admired; and they are part of the activity of discovery, of exploration, that is the activity of designing. Sketches by another architect, James Stirling (Figure 3.2), again show plans and sections being drawn and considered together, and also 3-D representations of the emerging building, as the designer generates alternative versions, and criticises them, and continues the voyage of discovery.

One might ask, why is it not done, simply to produce the one drawing, the one solution? Well, because a large number of alternative solutions are possible, and because (as Marples pointed out all that time ago) the nature of the design problem can only be found by exploring it through some alternative solution proposals.

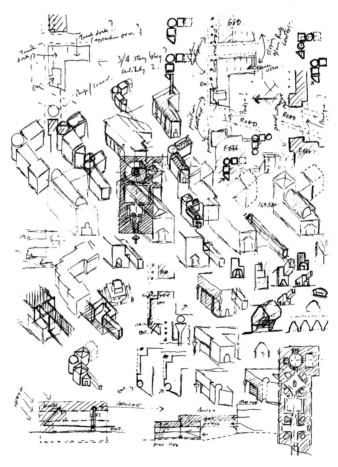

Figure 3.2. James Stirling: design sketches for the Fogg Museum, Boston, USA

This critical, reflective dialogue through sketching is just as relevant in high-performance engineering design as it is in architectural design. The racing car designer Gordon Murray says that he does lots of thumb-nail sketches, in which he 'talks' to himself by annotating them with criticisms such as 'too heavy', and 'stupid' or 'rubbish' (Cross and Clayburn Cross, 1996).

So what might we learn from looking at these examples of what designers sketch, and considering their own comments about why they make sketches? One thing that is clear is that sketches enable designers to *handle different levels of abstraction simultaneously*.

Clearly this is something important in the design process. We see that designers think about the overall concept and at the same time think about detailed aspects of the implementation of that concept. Obviously not *all* of the detailed aspects, because if they could do that, they could go straight to the final set of detailed drawings. So they use the sketch to identify and then to reflect upon critical details – details that they realise will hinder, or somehow significantly influence the final implementation of the detailed design. This implies that, although there is a hierarchical structure of decisions, from overall concept to details, designing is not a strictly hierarchical process; in the early stages of design, the designer moves freely between different levels of detail.

The identification of critical details is part of a more general facility that sketches provide, which is that they *enable identification and recall of relevant knowledge*. As Richard MacCormac said, 'What you need to know about the problem only becomes apparent as you're trying to solve it.' There is a massive amount of information that *may* be relevant, not only to *all* the possible solutions, but simply to *any* possible solution. And any possible solution in itself creates the unique circumstances in which these large bodies of information *interact*, probably in unique ways for any one possible solution. So these large amounts of information and knowledge need to be brought into play in a selective way, being selected only when they become relevant, as the designer considers the implications of the solution concept as it develops.

Because the design problem is itself ill-defined and ill-structured, another key feature of design sketches is that they *assist problem structuring through solution attempts*. We have seen that sketches incorporate not only drawings of tentative solution concepts but also numbers, symbols and texts, as the designer relates what he knows of the design problem to what is emerging as a solution. Sketching enables exploration of the problem space and the solution space to proceed together, assisting the designer to converge on a matching problem-solution pair. It enables exploration of constraints and requirements, in terms of both the limits and the possibilities of the problem and solution spaces.

Finally, as several design researchers have pointed out, sketches in design *promote the recognition of emergent features and properties* of the solution concept. They help the designer to make what Goel (1995) called 'lateral transformations' in the solution space: the creative shift to new alternatives. They assist in what Goldschmidt (1991) called the 'dialectics of sketching', the dialogue between 'seeing that' and 'seeing as', where 'seeing that' is reflective criticism and 'seeing as' is the analogical reasoning and reinterpretation of the sketch that, again, provokes creativity. And sketches help the designer to find the unintended consequences, the surprises that keep the exploration going in what Schön and Wiggins (1992) called the 'reflective conversation with the situation' that is characteristic of design thinking.

In all these ways, and in some more that I'm sure we have still to realise, sketching helps design thinking. In design, drawing is a kind of intelligence amplifier, just as writing is an intelligence amplifier for all of us when we are trying to reason something out. Without writing, it can be difficult to explore and resolve our own thoughts; without drawing, it is difficult for designers to explore and resolve their thoughts. Like writing, drawing is more than simply an external memory aid; it enables and promotes the kinds of thinking that are relevant to the particular cognitive tasks of design thinking. We have seen, through considering the role of sketching in design, confirmation of the aspects of design thinking I identified earlier, such as the exploratory, opportunistic and reflective nature of design thinking.

Can a Machine Design?

Let me now turn from understanding how human designers think, from natural intelligence in design, to artificial intelligence in design: from how humans think to how machines might think.

Asking 'Can a machine design?' is similar to asking 'Can a machine think?' The answer to the latter question seems to be, 'It all depends on what you mean by "think".' Alan Turing (1950) attempted to resolve the question by his 'Turing Test' for artificial intelligence – if you could not distinguish, in a blind test, between answers to your questions provided by either a human being or a machine, then the machine could be said to be exhibiting intelligent behaviour, *i.e.* 'thinking'.

In some of my research related to computers in design, I have used something like the Turing Test in reverse – getting human beings to respond to design tasks as though they were machines. There have been various intentions behind this strategy. One intention has been to simulate computer systems that do not yet exist; another has been to try to shed light on what it is that human designers do, by interpreting their behaviour as though they were computers. My assumption throughout has been that asking 'Can a machine design?' is an appropriate research strategy, not simply for trying to replace human design by machine design, but for better understanding the cognitive processes of human design activity. However, this assumption has been challenged recently. Here I will first review some of my research, and then return to this challenge.

My first postgraduate research project, at the Design Research Laboratory at UMIST, Manchester, directed by John Christopher Jones, was in 'Simulation of Computer Aided Design' (Cross, 1967). At its core was a novel but strange idea that we might get some insights into what CAD might be like, and what the design requirements for CAD systems might be, by attempting to simulate the use of CAD facilities which at that time were mostly hypotheses and suggestions for future systems that hardly anyone really knew how to begin to develop. The strangeness about this idea was that we would effect these simulations through getting human beings to pretend to be the computers! This was the reverse application of the 'Turing Test'.

The research was based on getting designers (architects) to attempt a small design project in experimental conditions (like the protocol studies and similar studies that have developed since that time). They were given the design brief, and asked to produce a sketch concept. As well as conventional drawing materials, they had a simulated computer system to help them: they could write questions on cards located in front of a closed-circuit TV camera, and would receive answers on a TV screen in front of them. In another room, at the other end of the CCTV link, was a small team of architects and building engineers who attempted to answer the designer's questions. Thus we had a very crude simulation of some features of what might actually now be parts of a modern-day CAD system, such as expert systems and databases.

The designers who participated in these experiments were not told what to expect from the 'computer', nor given any constraints on the kinds of facilities they might choose to ask of it. Thus we hoped to discover what kinds of facilities and features might be required of future CAD systems, and gain some insights into the 'systemic behavioural patterns' that might emerge in these future human-computer systems.

We conducted ten such experiments, each of which lasted about one hour. The messages between designer and 'computer' were recorded, and one of the analyses I made was to classify them into the topics to which they referred, from the client's brief to construction details. This kind of data gave some insight into the designers' patterns of activity, such as a cyclical pattern of topics over time, from requirements to details and back again. The number of messages sent in each experiment was quite low, with normally several minutes elapsing between requests from the designer. Of course, the response time from the 'computer' could also be quite long, typically of the order of 30 seconds. Despite this apparently easy pace of interaction, all the designers reported that they found the experiments hard work and stressful. They reported the main benefit of using the 'computer' as being an increased speed of work, principally by reducing uncertainty (*i.e.* they relatively quickly received answers to queries, which they accepted as reliable information).

We also tried a few variations from the standard experiments. The most interesting was to reverse the normal set of expectations of the functions of the designer and the 'computer'. The 'computer' was given the job of having to produce a design, to the satisfaction of the observing designer. It was immediately apparent that in this situation there was no stress on the designer – in fact, it became quite fun – and it was the 'computer' that found the experience to be hard work. This led me to suggest that CAD system designers should aim for 'a much more active role for the computer':

We should be moving towards giving the machine a sufficient degree of intelligent behaviour, and a corresponding increase in participation in the design process, to liberate the designer from routine procedures and to enhance his decision-making role. (Cross, 1967.)

This vision of the intelligent computer was based on an assumption that a machine *can* design – that it can be programmed to do a lot of the design work, but under the supervision of a human designer. I still think that there is something relevant in this vision of the computer as designer – it still offers a more satisfactory basis for the human-machine relationship in computer aided design than current CAD systems. Why isn't using a CAD system a more enjoyable, and perhaps also a more intellectually demanding experience than it has turned out to be?

Computation and Cognition

It seems that AI-in-design research can be aimed either at *supporting* design (through interactive systems that aid the designer's creativity) or at *emulating* design (that is, through developing computational machines that design). Where the goal is to develop interactive systems that support designers, then knowledge of the human designer's cognitive behaviour obviously is of fundamental importance, because the users of the interactive systems (that is, designers) must be able to use them in ways that are cognitively comfortable. So the systems must be designed on the basis of models of the cognitive behaviour of the system users.

But reference to human designers' cognitive behaviour is perhaps inevitable in AI research in design, because it is the results of human behaviour that set the standards for performance against which to assess the progress of the computational machines, just as chess-playing machines are matched against human chess masters. The highest standards of design performance that are (currently) available to us are those provided by the most creative human designers.

However, perhaps the goals of developing either interactive design systems or autonomous design machines are not the only goals of this research field. For some, an intrinsic goal is to further our understanding of human cognitive behaviour by attempting to model or emulate it in computational terms. Asking 'Can a machine design?' can be a research strategy for gaining insights into how designers think.

Again, we might make the analogy with computational modelling of chess playing. Surely the highest goals of such modelling are not simply to produce machines that can (in some sense) 'play' chess, or fatuously to replace all human chess players by machines? The goals of computational chess-playing must include developing our understanding of the nature of the 'problem' of the chess game itself, and of the nature of the human cognitive processes which are brought to bear in chess playing and in the resolution of chess problems. In such ways we could further our understanding of ourselves.

At least, that has always been my assumption about the value of having machines do things that human beings do for fun, whether that is playing chess, or designing. But John Casti, of the Santa Fe Institute, has come to a rather disturbing conclusion about the lessons that have been learned from chess-playing machines. In his book, *The Cambridge Quintet* (1998), Casti imagines some conversations on computation and artificial intelligence between Turing, Wittgenstein, Schrödinger, Haldane and Snow. In a postscript, Casti refers to the 1997 defeat of the World chess champion, Garry Kasparov, by the computer program Deep Blue II, and he quotes Kasparov as saying, 'I sensed an alien intelligence in the program.'

Casti then goes on to come to the rather surprising, and depressing conclusion that 'we have learned almost nothing about human cognitive capabilities and methods from the construction of chess-playing programs'.

So, in AI-in-design research, will we be forced to come to the same conclusion, that 'we have learned almost nothing about human cognitive capabilities and methods from the construction of designing programs'? Will designers rather nervously contemplate the 'alien intelligence' of the designing programs? Will we have built machines that can design, but also have to bring ourselves to Casti's view of the 'success' of chess-playing machines: 'the operation was a success – but the patient died!'?

Perhaps Casti is being unduly pessimistic. One thing that we have learned from chess-playing programs is that the brute force of computation actually can achieve performances that outmatch human performance in a significant area of human cognitive endeavour. And surely researchers of computer chess-playing have learned something of the cognitive strategies of human chess players, even though their programs do not 'think' like humans?

What is the point of having machines do things that human beings not only can do perfectly well themselves, but also actually enjoy doing? Like playing chess, people *enjoy* designing – and, I believe, people are *very good* at designing. If, as Casti says, the building of chess-playing computer programs really has taught us 'almost nothing about human cognitive capabilities', then the research and programming efforts have been almost useless.

It seems to me that the kind of AI research that emulates human cognitive activities should address the question, 'What are we learning from this research about how people think?' Some of the AI-in-design research should attempt to tell us something about how designers think. Through AI-in-design research, I think we can hope to learn some things about the nature of human design cognition through looking at design from the computational perspective.

And of course, there is something else as well. Instead of machines that do things that people enjoy doing, and are good at doing, we want machines to do things that are arduous and difficult for human beings to do. We also want machines to do things that are not merely arduous or difficult for human beings to do, but to do things that human beings simply cannot do unaided. So rather than just emulate human abilities, some of our design machines should also do things that designers cannot do.

4

Creative Cognition in Design I:
The Creative Leap

Literature on creativity often emphasises the 'flash of insight' by which a creative idea is frequently reported to occur. The classic accounts of creative breakthroughs in science and mathematics, such as Kekulé's account of his insight into the structure of the benzene molecule, or Poincaré's accounts of his mathematical insights, suggest that creative thought is characterised by such acts of sudden illumination (Koestler, 1964). Wallas (1926) incorporated this view into a general model of the process of creative problem-solving, which consists of the four stages of preparation, incubation, illumination and verification. This model is still accepted as valid today, and the concept of sudden 'illumination' as representing creative thought is so widely understood that cartoonists use a lighted lightbulb as a universal symbol for someone suddenly having a 'bright idea'.

Similarly, in engineering and design, significant innovations or novel design concepts are often reported as arising as sudden illuminations (Maccoby, 1991). The idea of 'the creative leap' has for some time been regarded as central to the design process (Archer, 1965). Whilst a 'creative leap' may not be a required feature of routine design, it must surely be a feature of non-routine, creative design. Some would argue that all design, by its very nature, is creative. However, there are times when a designer will generate a particularly novel design proposal, and there is evidence that the level of 'creativity' of a design proposal can be reliably assessed, at least by peer-groups (Amabile, 1982; Christiaans, 1992). In this case, creative design is related to product-creativity, rather than process-creativity.

In some other fields, the 'creative leap' is characterised as a sudden perception of a completely new perspective on the situation as previously understood. This is the basis of Koestler's (1964) model of 'bi-sociation' to explain the 'creative leap' for example in humour. In creative design, it is not necessary that such a radical shift of perspective has to occur in order to identify a 'creative leap'. There might

First published as 'Modelling the Creative Leap' in the preprints of the international workshop *Computational Models of Creative Design III*, edited by J S Gero, M L Maher and F Sudweeks, Key Centre of Design Computing, University of Sydney, Australia, 1995.

be no unexpected dislocation of the solution space itself, but merely a shift to a new part of the solution space, and the 'finding' there of an appropriate concept. This is what characterizes creative design as exploration, rather than search. Unlike bi-sociation, creative design is not necessarily the making of a sudden 'contrary' proposal, but is the making of an 'apposite' proposal. Once the proposal is made, it is seen to be an apposite response to the given, and explored, problem situation.

In this chapter, creative design is therefore regarded as the apposite proposal of a concept which embodies novel features for a new design product. Such a proposal may or may not arise as a sudden 'flash of insight', but it will constitute a 'creative leap' across the gap between the functional design requirements and the formal design structure of a potential new product. We shall see that the creative cognitive act in design appears to be not so much taking a leap as building a bridge between problem requirements and solution proposal.

An Example of a Creative Leap

This example of a 'creative leap' occurring in a design context comes from one of the protocol analysis studies used in the Delft Design Protocols Workshop (Cross *et al.*, 1996). This Workshop was based on a set of analyses made by different researchers around the world, of the same selected videotape recordings and transcripts of experimental design sessions. Two such experimental sessions were used in the Workshop; one using the 'think aloud' protocols of an individual designer, and the other using the naturally-occurring interactions of a small team of three designers (identified anonymously in the transcripts as I, J and K). The same design problem was set both to the individual designer and the team: the design of a carrying/fastening device for mounting and transporting a hiker's backpack on a mountain bicycle. This device would be something like a special bicycle luggage rack.

A 'creative leap' seems to have occurred as a sudden illumination in the team's design process, at a point when one of the team members, Designer J, suggests the following design concept: *'maybe it's like a little vacuum-formed tray'*. This tray idea is quite quickly taken up by the team, and the other members collaborate in developing the concept into a fully-fledged design. A transcript of the session at the period around the 'creative leap' is given in Appendix A. Their resulting design proposal is shown in Figure 4.1. The proposal is for a plastic tray, positioned over the rear wheel of the bicycle, with metal mounting points onto the saddle tube and onto the rear wheel frame tubes, and with cross-over straps to hold the backpack in the tray.

Records of their own work were kept by the team in the form of sketches on paper, and lists compiled on the whiteboard. They began by attempting to list a 'functional specification' and the problem constraints, and to this was added later a list of features that they intended their product to have. All these items were derived from the brief and related information provided in the experiment. They then developed the problem into three sub-problem areas: 1. the position of the rack device relative to the bicycle, 2. joining mechanisms between (a) the backpack and the rack and (b) the rack and the bicycle, 3. materials for making the rack.

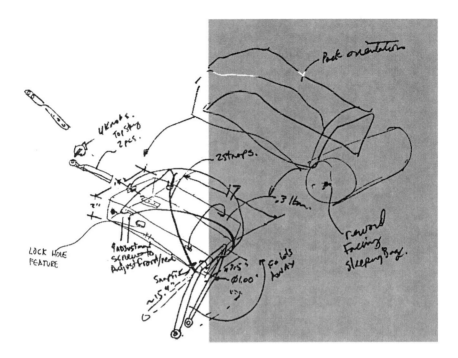

Figure 4.1. The team's design proposal

In each case, they explored problems and solutions together, by proposing concepts (sub-solutions) for each sub-problem and evaluating/discussing the implications and possibilities of each concept. For example, Appendix B is part of the transcript of the team's discussion of the 'rack-to-bike' joining problem. This shows how their thoughts about the positioning of the rack and its supports to the bicycle frame raise issues such as riding stability, ergonomics of use, weight of a full backpack, and user behaviour. In general, they argue from form to function, rather than *vice versa*.

One of the significant issues that arises in this way is that the backpack's own shoulder straps, *etc.*, become hazardous if they dangle into the bicycle wheel. After generating their random concept-lists, the team then review each list to eliminate unsatisfactory concepts and identify their preferred ones. As they go through the pack-to-rack list, the 'bag' concept is stressed as a solution for holding all the loose straps, and then the 'tray' concept suddenly appears, as the transcript shows:

I: Bag; put it in a bag; we're gonna need some sort of thing to do something with those straps

K: To get this out of the way

J: So it's either a bag, or maybe it's like a little vacuum-formed tray kinda for it to sit in

I: Yeah, a tray, that's right, OK

J: It would be nice, I mean just from a positioning standpoint, if we've got this (backpack) frame outline and we know that they're gonna stick with that, you can vacuum-form a tray

I: Right, or even just a small part of the tray...

K: Something to dress this [straps] in

J: Maybe the tray could have plastic snap features in it, so you just like snap your backpack down into it

K: Snap in these [backpack] rails

J: It's a multi-function part

K: You just snap in these rails

J: Yeah, snap the rails into the tray there

I: OK

J: It takes care of the rooster-tail problem ...

In this 1-minute segment, we see the key concept, the tray idea, being proposed, accepted, modified, developed and justified. As well as securely holding the backpack, the proposed concept solves two particular problems: the dangling straps problem and the 'rooster-tail' problem – *i.e.* the water/mud spray (like a rooster tail) thrown up by a mountain bicycle wheel, which would dirty the backpack unless it is protected. The conceptual strength of the tray idea seems to lie in the way it embodies a potential solution form that, once it has been expressed, recognisably satisfies certain key problems and also recognisably can be modified and refined to accommodate other problems and requirements in a satisfactory way. It is an 'apposite' proposal, as defined earlier.

But did the tray idea just come out of the blue? It was certainly the first instance of the use of the word 'tray' in the whole transcript, and from then on 'tray' is repeatedly used as the defining concept for the team's design proposal. (The word 'tray' subsequently occurs 35 times in the last 40 minutes of the transcript.) Possibly related concepts that had been mentioned earlier included references to injection-moulded plastic as a possible material, and flat plastic forms for the rack device. In fact, nearly 20 minutes earlier than the tray idea was first expressed, it's originator, Designer J, referred to a similar kind of rack idea that he recalled:

J: It looks like everything we're looking at right now is wire-form, but actually a friend of mine suggested a product that he would do – an injection-moulded rack that would kind of like fold down – a couple of years ago...

Another team member immediately responded with recalling a similar device that he remembered:

> I: It's like the little rack that was flat, it had these panels... but these panels were solid, it had little wheels... and it would come off and then it would be like a little trailer

Designer J also suggests another kind of flat plastic panel solution a few minutes later:

> J: I think that a super simple solution – might not be strong enough though - if you can imagine just taking a piece of like propylene or something like that, and diecutting this triangle that you can fold, you know, like a cutout from a pop-up book or whatever, and it bolts on down there, and creates a flat surface... kind of acts as a mudguard too

So ideas related to the device as a flat sheet of plastic, which would also act as a mudguard, were being suggested shortly before the appearance of the concept embodied in the apparent creative leap. The significant difference seems to be expressing this concept as a 'tray' – *i.e.* a flat surface with a raised lip around its circumference. (Proposing this as 'vacuum-formed' was also the first time this manufacturing process was mentioned, but as the concept is developed, the manufacturing process reverts to being injection-moulded.) The 'tray' concept summarises, in an envisionable form, a recognisably good solution, in a way that is significantly different from the potential concept of a 'flat', 'folded', 'panel'. The key difference seems to be related to perceiving a tray as a container (like a bag), whereas the previous concepts had only identified a flat surface.

As the earlier transcript extract showed, the first emergence of the tray concept seems to be immediately recognised and accepted by the team as a good concept. However, they return to their discipline of checking-off the other concepts that they had generated. But Designer J is careful to insist that the new concept of 'tray' is added to the list:

> J: I think tray is sorta, a new one on the list, it's not a sub-set of bag...

Very shortly afterwards, as they conclude this stage of their design process, Designer J also makes clear his commitment to the tray concept:

> J: I really like that tray idea ... I think all design eventually comes down to a popularity contest

The ways in which persuasive tactics are used by members of the team to get their own preferred concepts adopted, such as expressing emotional commitment to them, have been referred to in more depth elsewhere (Cross and Clayburn Cross,

1995). The emotional content of creative thought, in the context of computational modelling, has also been stressed by Gelernter (1994).

To summarise how this 'creative leap' emerged, we can see that it draws upon earlier notions that, in retrospect, seem very similar – a flat, folded surface in plastic material – but which lacked the apparently critical feature of 'containment' that a 'tray' has: its generation is perhaps aided by the immediately prior consideration of a more extreme form of containment, a bag; it seems to focus on one particular problem (containing the straps) as the most significant consideration; it is quickly elaborated to satisfy a range of other problems and functions; it is recognisably a bridging concept between problem and solution, which synthesizes and resolves a variety of goals and constraints; and it occurs during a 'review' period, after earlier periods of more deliberately generating concepts and ideas.

Identifying the Leap

The Delft Protocols Workshop was concerned with analysing design activity across a broad spectrum of approaches; it was not concerned specifically with analysing creativity, for example. Of the twenty Workshop papers, ten analysed in some form the team experiment, but none of these concentrated specifically on the 'creative leap' identified above. However, some of the analyses of design activity in these papers provide evidence which identifies when the 'leap' occurs, and its significance in the design process of the team.

Most analyses of the team design process in the Delft Workshop do not indicate how the tray concept originated, but some do reinforce the importance of this concept as marking a key point in the process. For example, Günther et al. (1996) classified the team's protocol statements into pertaining to three major stages of a design process: clarifying the task, searching for concepts and fixing the concept. Their resulting chart (Figure 4.2) suggests how the tray concept, which occurred at around 78 minutes, effectively ended the 'searching for concepts' stage. Similarly the graph produced by Maziloglou et al. (1996), of 'discourse production' (Figure 4.3), shows how the team's discourse (verbal statements made) peaked in the 'solution' related category in the period around the emergence of the tray concept. Radcliffe's (1996) analysis of the shifting 'work loci' (Figure 4.4) also shows how the focus shifts at around 80 minutes, from handling artefacts (principally the backpack and bicycle provided for the team) and listing on the whiteboard, to developing the final design, largely through sketches.

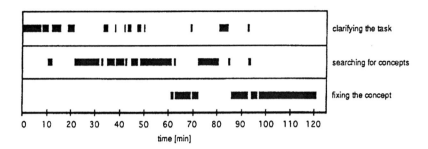

Figure 4.2. Principal phases of the team's design process, identified by Günther *et al.* (1996)

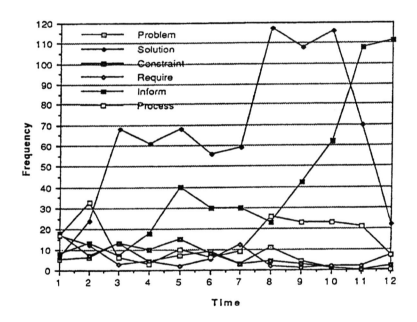

Figure 4.3. The team's discourse production over time (10-minute intervals), identified by Mazijoglou *et al.* (1996)

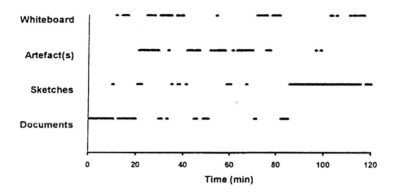

Figure 4.4. The team's shifting work loci over time, identified by Radcliffe (1996)

An analysis which came closer both to tracing the 'history' of the tray concept and indicating its important role was that by Goldschmidt (1996). Her 'linkograph' of the relevant section of the team protocol is shown in Figure 4.5, where J's 'tray' statement is number 30. The linkograph shows how each statement (or 'move') is linked (by a 'common sense' analysis of relationships between statements) to others. Statement 30 in this particular group is identified as a 'critical move', *i.e.* one which has a relatively high number of links to other statements that succeed it. Goldschmidt identifies this set of statements around statement 30 as a particularly 'productive' phase of the team's design activity, relatively rich in interlinks between statements. Again, her analysis does not explain how the significant 'tray' concept came to be generated, but her analysis confirms it as a statement that is very influential.

The linkograph shows a highly-interconnected 'chunk' of statements, from statements 28 to 54. Statement 28 is Designer I's suggestion to 'put it in a bag'; statement 54 is Designer J's insistence that 'tray is sorta, a new one on the list, it's not a sub-set of bag.' In that short period (2 minutes) we see that the tray concept somehow generates a highly productive, cognitively rich sequence of interacting statements, with the team members building on each other's ideas. (The full transcript of statements 28-54 is that given in Appendix A.)

Statement 30 ('maybe it's like a little vacuum-formed tray') does appear to come 'out of the blue' – it has just two 'back-links' in the linkograph, to the immediately preceding statements. (Other 'back-links', for example earlier references to flat, plastic devices, are not shown in this particular, limited section from the full session.) Its importance, though, is clear in the relatively high number of 'fore-links' it has; *i.e.* subsequent statements that build onto, or refer back to, this statement.

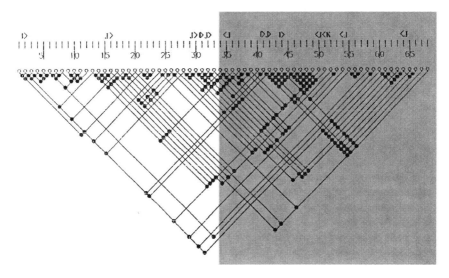

Figure 4.5. Linkograph of the team's design moves around the 'creative leap' (move 30), identified by Goldschmidt (1996)

Modelling the Leap

Research in artificial intelligence has attempted to model and simulate various aspects of design, including creative design. This section will draw from some of these attempts to model creative design, in order to see what insights they add to the previous example of a creative leap in the design process, and whether the example creative leap can be explained in computational modelling terms.

Rosenman and Gero (1993) and Gero (1994) suggested five procedures by which creative design might occur: combination, mutation, analogy, design from first principles, and emergence. In this section I will discuss how these procedures might be used to explain, or at least to shed some light on, the particular 'creative leap' example presented above. I will also discuss the possibilities and/or difficulties that there appear to be within computational modelling of providing adequate models of creative design through such procedures.

Combination

Creative design can occur by combining features from existing designs into a new combination or configuration. In the example of the 'tray' idea, relevant previous concepts that had occurred in the team's discussion were that of a flat plastic panel and a bag. It seems possible that the 'creative leap' occurred by a combination in the designer's mind of 'panel' + 'bag' to give 'tray' (Figure 4.6). In this case, 'tray' is not a new kind of artefact (trays already exist), but the combination of 'panel' + 'bag' in the designer's mind could have triggered an association with 'tray', as suggested by Figure 4.6. In the context of the team's design process, at that particular point, 'tray' was a novel concept.

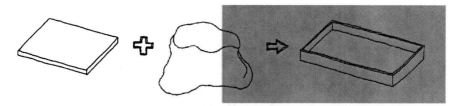

Figure 4.6. Possible combination of 'panel' plus 'bag' to give 'tray'

A more novel concept than a simple 'tray' might have arisen from the combination of 'panel' + 'bag'; for example, a bag with a normal, flexible upper part but a rigid, flat panel bottom (again, such artefacts do already exist). In fact, the team members do go on to propose developments of the tray idea which would have been more novel combinations of 'panel' + 'bag'. Immediately after the initial acceptance of the tray idea, Designer I articulates a concept of a net-like zippered container, which J develops into 'a tray with a net and a drawstring', and K (using analogy) further develops into the net as something like a retractable window blind:

> I: What if your bag were big, or, what if your, er, if this tray were not plastic, but like a big net, you just sorta like pulled it around and zipped there, I dunno
>
> J: Maybe it could be a part, maybe it could be a tray with a net and a drawstring on the top of it, I like that, that's a cool idea
>
> I: a tray with sort of just hanging down net, you can pull it around and zip it closed
>
> K: It could be like a window shade, so you can kinda, it sinks back in
>
> I: It retracts, yeah
>
> K: You pull down, it retracts in
>
> J: A retracting shade

In this sequence of the team's dialogue, we see how the initial (possible) combination of 'panel' + 'bag' ⇒ 'tray' becomes developed into a combination of 'bag' + 'tray' which leads to an original very novel concept of a tray with some form of retractable, net-bag container. (The lack of a familiar term to describe this device indicates its novelty.) This would have been perhaps a 'more creative' combination of 'panel' + 'bag' than the 'tray' concept. In the end, the team does not develop the retractable net-bag idea, but adds cross-over straps to the tray as a means of constraining the backpack.

The team seems to know how far to pursue novel combinations, before withdrawing to reconsider and start another line of reasoning. In computational systems it is difficult to know how to set such a limit; how does a system recognise that a satisfactory, or more-than-satisfactory concept has been created from combinations of previous concepts?

Mutation

Creative design by mutation involves modifying the form of some particular feature, or features, of an existing design. In computational systems, features may be selected and modified at random, and then evaluated, or there may be some directed procedure for selecting and modifying particular features. In the natural design process, it seems more likely that the latter procedure would operate.

In our example, a mutation procedure might conceivably have happened, transforming a flat panel into a tray (Figure 4.7). If Designer J was thinking of the inadequacies of a flat panel (*e.g.* it doesn't securely contain the backpack), he could have thought of putting a raised lip around the edges of the panel, giving rise to the concept of a tray. Designer K's earlier sketch (see Figure 4.10a, and the discussion of 'emergence', below) may also have been influential in suggesting such a mutation. We do not know what cognitive processes gave rise to J's 'creative leap', but it does seem that a mutation procedure could have generated 'tray' from 'flat panel'.

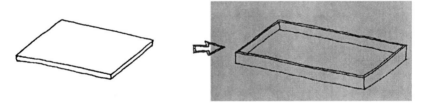

Figure 4.7. Possible mutation of 'flat panel' into 'tray'

The difficulty in computational modelling is identifying which structural features of the existing design to select for modification, and what kinds of modification to apply. In this case, to reproduce 'flat panel' ⇒ 'tray', it would have been necessary to identify the panel edges as relevant features, and to modify them by thickening and/or extending them out of the surface plane of the existing design. The mutation procedure would have to have been based on recognition of the inadequate behaviour of a flat panel in relation to the function of 'containment'.

Analogy

The use of analogical thinking has long been regarded and suggested as a basis for creative design (Gordon, 1961). We have already seen, in the extract above, how 'window shade' is used as an analogy to help describe (if not necessarily to generate) a concept of a retractable net-bag. The 'tray' idea does seem to originate in close association with the 'bag' idea. Designer J says, 'So it's either a bag or maybe it's like a little vacuum-formed tray, kinda, for it to sit in,' which suggests that he thinks of 'tray' as an alternative to 'bag' for the backpack to 'sit in'. This strongly suggests an analogical procedure 'bag' ⇒ 'tray' (Figure 4.8), based on thinking of analogues to 'bag' for something to 'sit in', to be contained and carried.

The difficulty for computational modelling based on analogy is in abstracting the appropriate behavioural features of an existing design. In this example, a bag's behavioural features of enclosing and carrying are apparently selected as relevant, whereas other behaviours such as flexibility are not. Furthermore, it would seem that partial-enclosure (such as in a tray) is more relevant than full-enclosure (as in a bag); about 20 minutes earlier in the session, before the 'tray' idea, J had suggested 'maybe it's a little bucket that it sits in,' but this was ignored by the rest of the team and apparently soon forgotten. 'Bucket' is more 'bag-like' than 'tray', but was apparently not deemed to be an appropriate analogy.

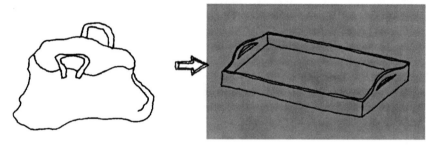

Figure 4.8. Possible analogy of 'bag' with 'tray'

First Principles

Designing 'from first principles' is often advocated as a way of generating good and/or creative designs (French, 1994). The difficulties for both artificial and natural design processes are in identifying what exactly the 'first principles' may be in any design situation, and how they may be used to generate design concepts. The example given by Rosenman and Gero (1993) is Peter Opsvik's design of the novel 'Balans' chair from the 'first principles' of the ergonomics of sitting posture. But what are the 'first principles' for 'a carrying/fastening device for mounting and transporting a hiker's backpack on a mountain bicycle'?

Perhaps we see an attempt at design from first principles in the sketch produced very early in the team's session by Designer K. This is reproduced as the left-hand side of Figure 4.9. K makes this sketch of 'backpack + accessory + bicycle' as though it is a personal attempt to represent the design problem – she does not draw it to the attention of the rest of the team, and it plays no overt role in the design process. However, perhaps it does express the 'first principles' of the design problem, and perhaps it does embody a 'tray-like' solution concept. Designer K later sketched such a solution concept, as discussed below.

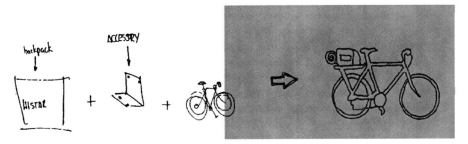

Figure 4.9. Possible inference of design from first principles

Designing 'from first principles' is at the core of any significant understanding of design. It assumes the theoretical position that designing proceeds by identifying requirements, or desired functions, and arguing from these to appropriate forms or structures. It is the abductive leap of reasoning from function to form that is regarded as the kernel of design (Roozenberg, 1993). But in practice, as we have seen in the extracts from the design team's protocols, and has been suggested by others (March, 1976), designers usually proceed by suggesting 'protomodels' of forms or structures, and evaluating these in order to amplify the requirements or desired functions. Takeda *et al.* (1996), in their analysis of the team protocol, showed how functions, as well as structures, develop and evolve during the course of the design process. The 'function' of a product to be designed is not, therefore, a static concept, a 'given' at the start of the design process.

Emergence

Emergence is the process by which new, previously unrecognised properties are perceived as lying within an existing design. Within the artificial intelligence community it has been discussed particularly with reference to the recognition of emergent, or extensional, shapes within the original, intentional shapes (Gero, 1994). However, emergent behaviours and functions, as well as emergent structures, are recognised by designers. For example, Designer J apparently recognises the emergent behaviour of protection from the 'rooster tail' spray in the tray concept, and adds that as a further validation of the concept.

In our example, it is difficult to know whether the 'tray' idea occurred as a case of emergence. In this context, it is interesting that Designer K had made a sketch quite early in the session (around 40 minutes) of what could be a design proposal which has a strong 'tray-like' resemblance (Figure 4.10a). As with her possible 'first principles' sketch, K does not publicly offer this sketch to the team, but makes the sketch whilst the other two team members are engaged in another activity. However, the other two certainly become aware of the sketch later, because they both use it (at around 60 minutes) to overdraw on it some different features – Designer J draws some adjustable stays onto it, and Designer I draws the wheels of his fold-down 'trailer' onto it. Designer I had just previously sketched the 'trailer' concept (Figure 4.10b).

Therefore it would be possible to speculate that 'tray' emerged as a structure from either Designer K's sketch or the previous concept of 'trailer' (Figure 4.10).

However, there is no real evidence for this. If it did, then the emergence procedure would seem to have been one of recognising the box-like structures in the sketches and converting that to a shallow box, *i.e.* a tray.

In anything other than flat-pattern, graphic or decorative design, emergence is not simply a matter of shape recognition. It involves recognising emergent behaviour out of structure, and/or emergent function out of behaviour. It therefore presents significant challenges in terms of computational modelling.

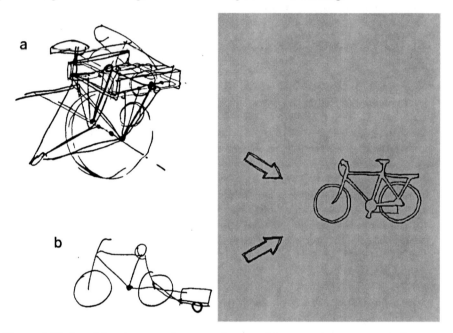

Figure 4.10. Possible inference of emergent concept from previous representations

Not Leaping but Bridging

This study of one example of a 'creative leap' in design has suggested that the example creative leap could conceivably be modelled by procedures such as combination, mutation, analogy, emergence, or designing from first principles. Because there is no overt record of the designers' cognitive processes, it is not possible to identify which, if any, of the creative procedures actually occurred in the example. However, if computational models of such procedures can be constructed, then progress is possible in computational modelling of creative design. Computational modelling of creative processes in the arts and sciences has had some reported success (Boden, 1990). The relative lack of progress in computational modelling of creative design may be due to the 'appositional' nature of design reasoning, in which function and form are developed in parallel, rather than in series.

The appositional nature of design reasoning has been neglected in most models of the design process. Consensus models of the engineering design process (Cross and Roozenburg, 1992), such as that promulgated by Verein Deutscher Ingenieure (VDI, 1987), the German professional engineers' association, propose that designing should proceed in a sequence of stages, like the stage-process adopted by the team studied here. They propose that a product design specification and a function structure should be developed before the search for solution principles and the generation of a principal solution. They also propose that the overall problem should be decomposed into sub-problems, and then sub-solutions found and combined into an overall solution. This is what the team attempted. However, as we have seen, exploration and identification of the complex network of sub-*problems* in practice is often pursued by considering possible sub-*solutions* (illustrated by the transcript extract in Appendix B).

In practice, designing seems to proceed by oscillating between sub-solution and sub-problem areas, as well as by decomposing the problem and combining sub-solutions.

During the design process, partial models of the problem and solution are constructed side-by-side, as it were. But the crucial factor, the 'creative leap', is the bridging of these two partial models by the articulation of a concept (the 'tray' idea in this example) which enables the partial models to be mapped onto each other. The 'creative leap' is not so much a leap across the chasm between analysis and synthesis, as the throwing of a bridge across the chasm between problem and solution. The 'bridge' recognisably embodies satisfactory relationships between problem and solution. It is the recognition of a satisfactory concept that provides the 'illumination' of the creative 'flash of insight'.

This recognition is a perceptual act by the designer (and by his colleagues, in this example of teamwork), and our knowledge of perceptual 'puzzles' can perhaps provide analogies of the process. For example, the recognition of a proposed design concept as embodying both problem and solution together may be regarded as something like the well-known duck-rabbit puzzle (Figure 4.11); it is neither one nor the other, but a combination which resolves both together and allows either to be focused upon. Suggesting that 'Maybe it's a little vacuum-formed tray' is rather like saying 'Maybe it's a duck-rabbit'. Can computational models of creative design recognise a duck-rabbit when they see one?

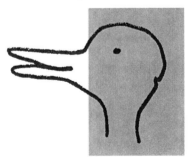

Figure 4.11. The duck-rabbit perceptual puzzle

Appendix A: Transcript of the discussion at the time of the 'creative leap'

I   we'll just call it that for now er bag put it in a bag we're gonna need some sort of thing to do something with those straps

K   to get this out of the way

J   yeah

I   yeah either the

J   so it's either a bag or maybe it's like a little vacuum formed tray kinda for it to sit in

01:19:00

I   yeah a tray that's right OK

J   'cos it would be nice I think I mean just from a positioning standpoint if we've got this frame outline and we know that they're gonna stick with that you can vacuum form a a tray or a

I   right or even just a small part of the tray or I guess they have these

K   so something to dress this in

J   yeah

I   or even just em

J   maybe the tray could have plastic snap features in it so you just like kkkkkk snap your backpack down in it

I   mmmm I was thinking of er

K   snap in these rails

J   it's a multifunction part huh

K   you just snap in these rails

J   yeah snap the rails into the tray there

K   mm mm

I   OK

J   it takes care of the easy it takes care of the rooster tail problem on your pack

I   uh uh what if your bag were big er what if you're you're on er in this tray were not plastic but like a big net you just sorta like pulled it around and zipped there I dunno

J   maybe it could be part maybe it could be a tray with a with a net and a drawstring on the top of it I like that

I   yeah I mean em

J   that's a cool idea

I   a tray with sort of just hanging down net

01:20:00

you can pull it around and and zip it closed

K   it could be like a a a window shade so you can kinda it sinks back in so it just

J   oh yeah

I   it retracts yeah

K   you pull down it retracts in

J   a retracting shade

I   right right

K   so that that's not dragging in the spokes if you don't have anything attached

J   so what we're doing right now though is we're coming up with like again classifications of solutions of kind of all they're all either or things I mean like we wouldn't do the net and the shade and the snap in with the tray either or any one of those will probably

I   yeah OK

K   a net can be combined with a shade I mean you could have a retractable net that that's how I thought of it

I   so we I think the issue that we're talking about is is straps so we'll just keep that one on the burner

K   yeah maybe there's some cool innovation there

J   well yeah OK

I   OK now er it had er has er

J   I think tray is sorta a new one on the list it's not a subset of bag it's a kind of er yeah but oh yeah yeah yeah oh I see shade straps is how do you dress the straps on the back

I   yeah yeah OK

Appendix B: Transcript of part of the discussion of joining 'rack to bike'

J   um one of the things I was thinking that if you did this one of the things that could be neat is people were talking about like centre of gravity and I think that it'll be

00:46:00

different for different people what their preference is a little bit

K   mm mm

J   like where they want that mass maybe the if there's a thing that comes down to here you could have it so that it adjusts so you could kinda lever the pack up or down a little bit y'know if it's not a a fixed

K   seems like lower is better regardless as you say like we design in the low position and not necessarily try and get

I   you're gonna have um

K   the adjustability

I   is there gonna be an issue of the height of this I mean

J   what about clipping under the bottom of the seat

I   yeah or even the the seat post neck

J   oh these things yeah

K   the other thing we ought to be concerned about ergonomically is that when you're at the bottom of your stroke your leg is is right in here you want to make sure you don't get too close to the seat

J   so you need to you need to come back from that

I   or

K   not too far back

I   lower back

K   yeah it's just one thing I've noticed when I put stuff on a big bike rack and it's sticking out kinda like a tent back here

I   yeah it'll bend to your legs yeah

K   then the back of my legs I can feel it

J   I mean what what how much weight do you think somebody could realistically put in that pack

K   probably thirty fifty

00:47:00

thirty pounds

J   fill it with sand

I   is that information we have access to um

J   yeah what's typical weight that people carry in a backpack

I   do we have information about what er weights are that people might carry in a backpack or

J   have they done any market surveys

I   market surveys about

Experimenter   We do have some facts on the use of the backpack here

J   OK – fiftyfive and sixtyfive litre versions of the backpack

K   twentytwo kilograms

J   so fortyfive pounds fifty pound yeah

K   including sleeping bag oh so I suppose that's an issue too when you put this thing on

I   oh yeah

K   you want to make sure that that is

J   still fits

I   (inaudible)

J   it says people are generally going to put that at the base of the pack

5

Creative Cognition in Design II:
Creative Strategies

Most studies of designer behaviour have been based on novices (*e.g.* students) or, at best, designers of relatively modest talents. The reason for this is obvious – it is easier to obtain such people as subjects for study. However, if studies of designer behaviour are limited to studies of rather inexpert designers, then it is also obvious that our understanding of expertise in design will also be limited. In order to understand expertise in design, we must study expert designers. In some instances, it will be necessary to study outstanding, or exceptionally good designers. This is analogous to studying chess masters, rather than chess novices, in order to gain insight of the cognitive strategies and the nature of expertise in chess playing.

As in chess playing, in design practice it also seems clear that some individuals have differing levels of ability – some designers seem to perform consistently better than others, and some are outstandingly good. However, there have been only a few studies of outstanding designers, such as Lawson's (1994) studies of successful architects and Roy's (1993) studies of successful product designers.

This chapter reports three studies of innovative design by outstanding designers – a protocol study and two retrospective case studies. The studies are of projects by the engineering designer Victor Scheinman, the product designer Kenneth Grange, and the racing car designer Gordon Murray. The focus is on identifying the designers' creative strategies in responding to the problems they tackle. There appear to be several striking similarities in the creative strategies exercised by these designers in these projects, which suggest that a common understanding, and indeed a general model might be constructed of high level, creative strategies in design.

First published as 'Strategic Knowledge Exercised by Outstanding Designers' in the preprints of the international workshop *Strategic Knowledge and Concept Formation III*, edited by J S Gero and K Hori, Key Centre of Design Computing, University of Sydney, Australia, 2001.

Studies of Outstanding Designers

Victor Scheinman

Victor Scheinman is an engineering designer with many years of experience in designing both mechanical and electro-mechanical machines, and robotic systems and devices. He was one of the earliest designers of modern robotic devices, and he has won several design awards from the American Society of Mechanical Engineers. He is an accomplished, expert designer, outstanding in his field. He volunteered to participate in a protocol study experiment, in which he was videorecorded whilst he 'thought aloud' over a 2-hour session. The observations of Victor's design strategy are therefore based on the artificial situation of a controlled experiment. Although Victor has a wealth of design experience, the design task set in the experiment was a novel task for him. The task was to design 'a carrying/fastening device that would enable you to fasten and carry a backpack on a mountain bicycle'. Full details of the experiment are reported in the proceedings of the 'Delft Design Protocols Workshop' (Cross *et al.*, 1996).

In the following analysis of Victor's strategy, quotations are taken from the transcript of his 'think aloud' comments, preceded by the timestamp for the quotation. After some preliminaries, the substantive experimental session began at timestamp 00.15 minutes.

Quite early in the session Victor began to identify particular features of the problem that would influence his approach to developing a design concept. For example, very early in the session, in reading the design brief, he made a comment that suggested he saw something special about the design problem:

> (00.19) it is to attach to a bicycle, a mountain bike, and to me that makes it different.

Victor was also able to draw on personal experience that helped him to formulate some of the implicit requirements for a good design solution:

> (00.26) having used a backpack on a bike in the past and having ridden over many mountains, unfortunately not on a mountain bike but I can imagine that the situation is similar, I learned very early on that you want to keep it as low as possible.

He also drew upon personal experience to confirm that the preferred location for the backpack would be on the rear wheel rather than the front wheel:

> (00.51) my first thought is hey the place to put it is back here; there's another advantage by the way of having it in the back I can see immediately, and that is it's off the side in the front, and you're on a mountain bike trail and you hit something you're out of control in the front wheel.

(00.52) downhill work on mountain bikes, I know you want to keep your weight back rather than forwards.

Victor's personal experience of biking with a backpack led him to identify an issue that only someone who has had such experience might be aware of:

(00.55) when I biked around Hawaii as a kid that's how I mounted my backpack ... and I have to admit if there's any weight up here this thing does a bit of wobbling, and I remember that as an issue.

So the view that Victor formed of the problem was that of the total task that encompasses the dynamic system of the rider plus bicycle plus backpack, and the issues of control of the bicycle that arise in the situation of riding over rough terrain with a heavy backpack attached to the bicycle. This is a different situation to that of everyday, smooth-surface, level-grade riding, and it accentuates the needs to position the backpack low and to the rear. The view that Victor had of the design task was significantly different from a view that might be formed from considering the bicycle and backpack in a static situation, or without considering the effects on the rider's ability to control the bicycle with a mounted backpack. Victor's understanding of the dynamic situation therefore enabled him to formulate a broad view of the design task.

From this overview of the total dynamic system of rider + bicycle + backpack, Victor identified stability as a key issue. Quite early in the session, commenting on a prototype design that had been developed earlier by other designers, he surmised about the user-evaluation report on this prototype that:

(00.22) it probably says the backpack's too high or something like that, and that bicycle stability's an issue.

Victor therefore framed the problem as 'how to maintain stability', given that a heavy backpack had to be carried over the rear wheel of the bicycle, and given his experience of the 'wobbling' that can occur in the riding situation. This problem framing and his prior experience led him to conclude that he must design a rigid carrying device:

(00.59) the biggest thing that I remember in backpack mounting is that it's got to be rigid, very rigid.

He then developed this viewpoint into the requirement that the structural members of any carrying device must be stiff:

(01.06) making the carrier stiff enough for holding the backpack, that seems to be a big issue.

So, at about halfway through the session, Victor had derived a framing of the problem which directed him to design a stiff, rigid carrier, mounted as low as possible over the rear wheel. Soon after, a secondary viewpoint emerged, which arose from considering the client's needs as well as those of the user (which had dominated Victor's thinking so far). The client for the design task was a manufacturer who wanted to sell the carrying device in conjunction with their already-existing backpack. The device therefore needed to have unique selling points that differentiated it from other, similar products. During the development of his design concept, Victor kept in mind that he needed the product to have a 'proprietary feature', as emerged in some of his comments, discussed below.

Having established a need for rigidity, Victor was able to utilise his knowledge of structural engineering principles as he developed a concept design for the carrying device – in particular, knowledge that a triangulated structure is inherently rigid. This led him to avoid designing a rectangular, parallelogram form of structure, which was the form that rather naturally seemed to arise from considering the basic shape of the carrier and the location of its supporting structure on the bicycle. Whilst sketching a basic position and layout for the device, Victor commented:

> (01.07) one of the problems with a bicycle carrier where the frame is mounted out here and it goes to that, is that you end up with a parallelogram – bad thing, bad thing!

He expanded on this comment, identifying his concern with stability as a key requirement:

> (01.08) if I were to make a frame that looked like this, that would be a very poor design because basically what I've got is, I've got a parallelogram which has very little lateral stability.

He then introduced the principle of triangulation, whilst drawing a triangular form onto the layout:

> (01.09) it would be nice if I could, for instance, run these rods up here to some point and therefore create a triangle, this would give me great stiffness – good idea!

The principle of triangulation subsequently guided Victor's generation of the basic form and the detailed design features of his carrier. As he drew his design in more detail, he commented:

> (01.16) we're going to have this as a triangular structure here to provide the lateral stability.

As he continued to develop his design (Figure 5.1), he constantly referred to structural principles, seeking to avoid 'bad' configurations and to generate 'good' ones, making comments such as:

> (01.42) my detail here is going to have to be something like this because my forces along this tube are this way . . good, this is good; and then this

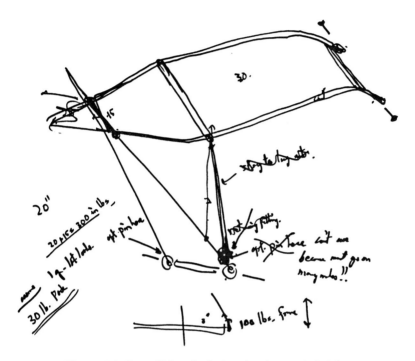

Figure 5.1. One of Victor's design development sketches

detail is going to be, er, let's see . . alright that's bad . . that's bad . . that's bad, so I'm going to need something like that.

In the meanwhile, as noted above, Victor also used the client's requirement of a unique selling proposition to help guide and to reinforce his decision to seek a design based on triangular structures:

(01.10) that is going to be our proprietary feature, a triangular, rigid structure with no bends in it; these rods are then going to be in tension and compression, no bending.

(01.41) I want to make sure that this rod here comes to a point, not stop right there . . that's to a point; that's going to be my feature.

In these comments, Victor demonstrated that he regarded the pronounced triangular form at the rear of the carrier as something to be retained as a feature that would help give the product an attractive and distinctive appearance. His design for the carrying device is therefore based on an integrated concept in which user requirements are addressed through the problem frame of stability, leading to the use of triangulation as the guiding first principle, which he then also uses to address the client's goal of having a proprietary, unique selling feature to the product.

Having the benefit of a 'think aloud' transcript, we can see that Victor's creative strategy involves addressing issues at several levels of generality – forming a broad, system's view of the required product in its situation; from that, developing a particular perspective or problem frame for guiding the solution concept; using that perspective to identify relevant first principles of engineering design to embody the concept; and also maintaining in mind the satisfaction of the client's goal of a successful consumer product. Let's now see if this analysis of a formal experiment helps to identify whether similar strategies can be observed in the real-world work of other outstanding designers.

Kenneth Grange

Kenneth Grange is a highly successful British designer of a great variety of products that range in scale from ball-point pens and disposable razors to train seats and railway engines. His career has spanned over more than forty years, and many of his designs became (and remain) familiar items in the household or on the street – or on the rail track. These designs include the first UK parking meters for Venner, food mixers for Kenwood, razors for Wilkinson Sword, cameras for Kodak, typewriters for Imperial, clothes irons for Morphy Richards, cigarette lighters for Ronson, washing machines for Bendix, pens for Parker, and the front end – driving cab and nose cone – of the British Rail high-speed train. He is one of the Royal Society of Arts' élite corps of 'Royal Designers for Industry', and his designs have won many awards. In 2001 he was awarded the Prince Philip Designer Prize – a kind of 'Oscar' for lifetime achievement. His career began with his first independent commissions in the nineteen-fifties, and in 1972 he was a founding-partner in what was to become the world-renowned interdisciplinary design consultancy, Pentagram. This study is based on personal conversations and a more formal taped interview with Kenneth (a fuller presentation of the study has been given elsewhere, in Cross, 2001).

A significant feature of much of Kenneth Grange's design work is that it is not based on just the styling or re-styling of a product. His designs often arise from a fundamental reassessment of the purpose, function and use of the product. A typical example is his design of a sewing machine for the Japanese company, Maruzen. They were looking for new designs for their European market where their high quality, well-engineered machines were sold under the name of Frister & Rossman. Kenneth's resulting design incorporated the standard Maruzen machinery, but repackaged it in novel ways that made it easier to use and gave the overall machine a new and distinctive form and style, as required by the clients.

The origins of the new design features lay in Kenneth's functional, practical approach, and on his personal experience. His starting point was to make some personal use of a sewing machine. He quickly found what he regarded as a 'contradiction' in the design, in that the sewing machine mechanism was located centrally on its base, whereas the user needs more surface space on their side of the needle than behind it. The user needs to assemble and lay out the material to be sewn, and control it as it passes under the needle, and therefore needs a flat working surface in front of the needle; once the work is behind the needle there is not the same need for space. Kenneth therefore saw one important aspect of the problem as being to increase the available work-surface space in front of the sewing needle. His solution concept simply moved the sewing machine mechanism rearwards on its base, creating an asymmetrical layout with more base-table space in front of the needle than behind it. To him, this appeared a virtually self-evident improvement to make, but the reason it had not been done before was because sewing machines had developed from regular engineering practices. The mechanism was from the very beginning simply put centrally on the base and nobody had thought about challenging this arrangement.

Another radical change in this particular sewing machine design was also a result of a simple, fundamental assessment of how the machine is used. Kenneth gave the base of the machine pronounced, rounded lower edges, which look like a mere styling feature, but in fact also arose from function. A recurring need in using a sewing machine is to clean the bobbin mechanism (under the needle, in the base of the machine) of the lint and loose fibres that inevitably gather and affect the functioning of the machine. In previous designs, this was achieved by the user tilting the machine backwards, away from them, into a precarious, unstable position that only allowed restricted access to the shuttle mechanism. To Kenneth Grange, this was simply inadequate. He wanted the user to be able to get easy, unrestricted access to the mechanism. So he designed it to tilt upright to the side, and that action in itself suggested a rolled edge to the base plate. The rolled edge made it easier for the user to tilt the machine, it rested stable and secure, and the complete underside was accessible for cleaning and oiling the lower mechanisms. A radiused top front edge was also provided to the base plate, to allow the fabric to slide over it more easily, and various other features were added, such as small drawers for holding accessories.

The sewing machine design illustrates how Kenneth Grange's approach is to consider the whole pattern of use of the product he is designing, exemplified here by considering the requirements of periodically cleaning the machine, and by considering how the user prepares and introduces the fabric into the stitching mechanism, thus requiring more make-up space in front of the needle than behind it. The innovative 'style' and features of the new machine were generated from considering and responding to the functional patterns of its use.

Gordon Murray

This case study is of a designer with a long-established record as a highly successful and highly innovative designer in a highly competitive environment; that of Formula One racing car design. Gordon Murray was chief designer for the Brabham team from 1973 to 1987, and the McLaren team from 1987 to 1991. Brabham cars designed by him and driven by Nelson Piquet won World Championships in 1981 and 1983, and his McLaren cars, driven by Alain Prost and Ayrton Senna, won World Championships in 1989 and 1990. In over 20 years in Formula One design, he established an outstanding reputation not only as a successful designer (over 50 race wins) but also as a consistently radical innovator.

Gordon Murray is clearly an outstanding designer who achieved considerable success. In his case the measures of his success as a designer are absolute – his achievements have been in a competition field where absolute performance standards are the criteria. We have been able to gain some insight into Gordon Murray's design strategies and approaches through conversations and interviews, and these have been reported in more depth elsewhere (Cross and Clayburn Cross, 1996). Here we will discuss just one example of Gordon's radical approach to racing car design.

At the start of the 1981 racing season, the Formula One governing body, FISA, had introduced new regulations intended to reduce 'ground effect' on the cars. This effect had been pioneered on Lotus cars some three seasons earlier; very low, smooth underbodies, flexible side-skirts and careful aerodynamic design provided a ground-effect downforce which increased the car's grip on the track surface. This meant much higher cornering speeds were possible, and by the 1980 season there were worries about safety and the lateral g-force effects that were being imposed on the drivers. In 1981 FISA set a minimum ground clearance under all cars of 6 cm, by which they intended to eliminate or substantially reduce 'ground effect'. But for Gordon Murray this change in the regulations was simply a stimulus to innovation. He said, 'The 1981 car, which was a World Championship-winning car, came absolutely from the regulation change. You sit there and you read the regulations and think, how we are going to do it? How the hell can we get ground effect back?'

Gordon realised that the authorities had to accept that at some points during a race, any car's ground clearance is going to be less than the 6 cm minimum, simply because of the effects of braking, or roll on corners, *etc*. His radical solution concept – which he said came as a sudden illumination after a long period of worrying at the problem – took advantage of this. Knowing that any driver-operated, mechanical device to alter the ground clearance was illegal, he focused on the physical forces that act on a car in motion. The braking and cornering forces he felt unable to work with because of their asymmetrical effects on the car, but the downforce pressure from airflow over a fast moving car will, if the car is well designed aerodynamically, push the car down equally over its whole length and width. The design challenge, therefore, as Gordon interpreted it, was to let the natural downforce push the car down at speed, and then somehow to keep it down when it slowed for corners, but allow the car to return to 6 cm ground clearance at standstill. Gordon had therefore framed the problem as one of sustaining a temporary lowering of the car, from natural forces, only whilst it was at racing speeds.

The ingenious solution that he developed incorporated hydro-pneumatic suspension struts at each wheel, connected to hydraulic fluid reservoirs. As the car went faster, the aerodynamic downforce pushed the body lower on its suspension and the hydraulic fluid in each suspension strut was pushed out into the reservoirs. The trick then was to find a way of letting the fluid return to the suspension struts only very slowly when the car slowed down. At cornering speeds, the suspension would stay low, but on slowing down and stopping at the end of the race, the fluid would return from the reservoirs to the suspension struts, giving the required 6 cm ground clearance. Gordon and his team developed such a system, using devices such as micro-filters borrowed from medical technology. The hydro-pneumatic suspension system is an example of radical innovation arising through framing the problem in a particularly focused way and then working creatively with basic physical forces.

Comparing the Strategies

Although they stem from very different domains of design – a bicycle luggage carrier, a sewing machine, a racing car – all three studies can be seen to demonstrate similarities in the approaches taken by the designers. Firstly, all three designers either explicitly or implicitly rely upon 'first principles' in both the origination of their concepts and in the detailed development of those concepts. Victor Scheinman relied strongly on the basic structural principle of triangulation to achieve the rigidity and stiffness that he considered important in the design of the backpack carrier. Gordon Murray stressed the need to 'keep looking at fundamental physical principles' for innovative design, and in his design to regain ground effect he focused on the physical forces that act on a car at speed. Kenneth Grange was less explicit about first principles, but it is clear that he adheres strongly to the modernist design principle of 'form follows function'; he approaches design problems 'by trying to sort out just the functionality, just the handling of it, and by-and-large out of that comes a direction.' This approach is evident in the sewing machine design, which is based very much on functional, usability aspects. So use of 'first principles' seems to be a crucial aspect in the knowledge and skills exercised by these three designers.

Secondly, all three designers appear to explore the problem space from a particular perspective in order to frame the problem in a way that stimulates and pre-structures the emergence of design concepts. In some cases, this perspective is a personal one that they seem to bring to most of their designing. For example, Kenneth Grange has a strong, emotional distaste for what he considers to be 'contradictions' in design, where the object is not well-adapted to its user and the patterns of use. He said, 'I think it's a question of what your attitude is towards anything, any working thing. My attitude is to want it to be a pleasure to operate.' And it was from operating the sewing machine that the essential concept of an asymmetrical layout emerged, and the rounded edges, which gave the clients the re-styling that they wanted. Victor Scheinman also used a distinct usability perspective in his problem structuring for the backpack carrier, for which, like Kenneth Grange, he drew upon his personal experience of using such a device. For Victor, it soon emerged that 'bicycle stability's an issue', and so 'it's got to be rigid, very rigid.' This led him to the triangularity of his design concept, which he then used to establish a distinctive appearance for the product, to satisfy the client's need for a unique selling feature. In both the sewing machine and the backpack carrier examples we see how the designer's personal problem framing and use of first principles led to a concept that reconciled the designers' goals (on behalf of the user) with the more commercial goals of the client. In the case of the racing car design, Gordon Murray's problem frame was governed by his focus on 'How the hell can we get ground effect back?' in order to achieve his goal of the fastest car, whilst satisfying the criteria set by the FISA regulations. This problem frame, and reliance upon first principles of 'basic physics', led him to the unique concept of the variable hydro-pneumatic suspension system. For these three designers, therefore, their problem framing arises from the requirements of the particular design situation, but is strongly influenced by their personal motivations, whether they may be altruistically providing pleasure for the product user, or competitively achieving the fastest car despite the regulations.

Finally, it seems from these three examples that perhaps creative design arises especially when there is a conflict to be resolved between the (designer's) high-level problem goals and the (client's) criteria for an acceptable solution. Such a conflict is particularly evident in Gordon Murray's design strategy, which was to challenge and somehow circumvent the criteria set by the technical regulations. In Kenneth Grange's case, the potential conflict was with the client's criteria for a product re-styling job, whereas his goal was to provide the user with an enhanced affordance of use from the product. As he said, 'You are almost invariably brought in by somebody who has got a very elementary commercial motive in changing the perception of the product. It's extremely unusual to be brought in to approach it from this usability, this function theme.' A very similar conflict was resolved by Victor Scheinman, when he reconciled the user's need for a stable, rigid product, with the client's commercial need for a product that had some distinctive marketing feature.

These similarities in strategy are illustrated in Figure 5.1. In each case, at the upper level there is a conflict, or potential conflict, between what the designer seeks to achieve as the highest goal and what the client sets as fundamental criteria. At the intermediate level, the designer frames the problem in a personal way, and develops a solution concept to both match that frame and satisfy the criteria. At the lower level, all three designers use first principles of basic physics, engineering and design to bridge between the problem frame and a solution concept.

A model that encapsulates and generalises the particulars of all three examples is given in Figure 5.2. At the lowest level the designer draws upon explicit, articulated knowledge of first principles, which may be domain specific or more general scientific knowledge. At the intermediate level is where strategic process knowledge is especially exercised, and where that knowledge is more variable, situated in the particular problem and its context, tacit and perhaps personalised and idiosyncratic. At the higher level there is a mix of relatively stable, but usually implicit goals held by the designer, the temporary problem goals, and fixed, explicit solution criteria specified by the client or other domain authority.

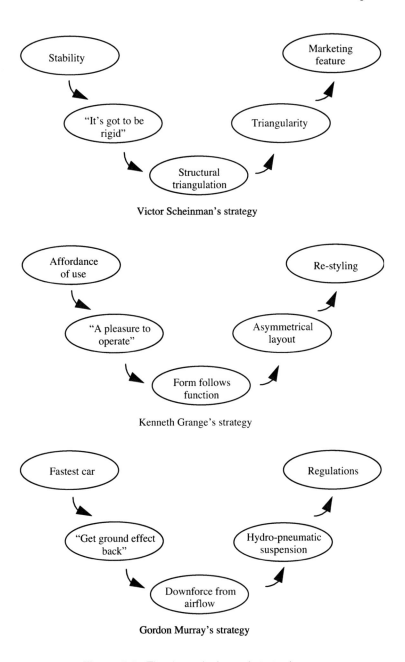

Figure 5.1. The three designers' strategies

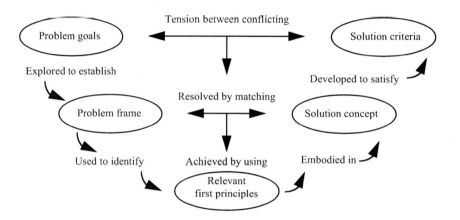

Figure 5.2. A general model of the creative strategy followed by all three designers

Design Expertise

Three key strategic aspects appear to be common in the creative expertise of all three designers: 1) taking a broad 'systems approach' to the problem, rather than accepting narrow problem criteria; 2) 'framing' the problem in a distinctive and sometimes rather personal way; and 3) designing from 'first principles'. These are all aspects that have been (separately) recommended by design theorists or methodologists from time to time. For instance, Jones (1970) has recommended a systems approach; Schön (1983) has identified the importance of 'problem framing'; and authors such as French (1985) and Pahl and Beitz (1984) have stressed the importance of 'first principles' as design guides in engineering. However, such insights and recommendations have not, in general, been based on much apparent evidence or empirical study of expert designers. The studies presented here therefore lend some credence to such insights, and perhaps offer more objective evidence about the nature of skilled, expert design behaviour.

The general model presented in Figure 5.2 also attempts to integrate the separate insights, and perhaps offers a broader understanding of outstanding, creative thinking in design. For example, although use of 'first principles' is often stressed in design education and practice, it is not evident *which* particular first principles are relevant to call upon until a problem frame has been established. The model, and the examples, perhaps also help to articulate some more detailed features, in the context of design, of the general observation often made by others (Ericsson and Smith, 1991) of the 'breadth-first' approach of experts in comparison with the 'depth-first' approach of novices.

From the analysis of the three examples, it appears that there are similar aspects to the creative strategies adopted by all three exceptional designers. It is perhaps surprising to see such commonalities between the three, considering the great disparity between the design projects in which they were engaged. However, although there are similarities in creative strategies across domains, this does not necessarily mean that experts can successfully switch practice between domains. Ericsson and Lehmann (1996) found that the superior performance of experts is usually domain-specific, and does not transfer across domains. Extensive training within a domain still seems to be crucial to professional expertise.

It is also worth commenting on the fact that the similarities in creative strategies found in these studies emerged from two quite different kinds of study – retrospective interviews and a concurrent protocol analysis. Nevertheless, there remain methodological problems of verifying the accuracy or relevance of the analyses that we and others have so far been able to make of the skills of exceptional designers. The difficulties of studying the performance of such people in formal ways may always limit the validity of the analyses, but more studies of expert and exceptional designers might lead to a more informed consensus about how design skills are exercised by experts, and on the nature of expertise in design.

6

Understanding Design Cognition

In this chapter I will focus on what we have learned about design cognition from protocol and other empirical studies of design activity. I will try to pick out some consistent patterns that may be discerned in the results of such studies, and to identify issues that are pertinent to the utilisation of the results (for example, in design education) and to further research. I will take a cross-disciplinary, or domain-independent view of the field, and try to integrate results from studies across the various domains of professional design practice.

Of all the empirical research methods for the study and analysis of design activity, protocol analysis (Ericsson and Simon, 1993) is the one that has received the most use and attention in recent years (Cross *et al.*, 1996). It has become regarded as the most likely method (perhaps the only method) to bring out into the open the somewhat mysterious cognitive abilities of designers. It does, though, have some severe limitations, to be noted later.

In analysing design cognition, it has been normal until relatively recently to use language and concepts from cognitive science studies of problem solving behaviour. However, it has become clear that designing is not normal 'problem solving'. We therefore need to establish appropriate concepts for the analysis and discussion of design cognition. For example, designing involves 'finding' appropriate problems, as well as 'solving' them, and includes substantial activity in problem structuring and formulating, rather than merely accepting the 'problem as given'. The first main area in which I will present my interpretations of findings, patterns and issues in design cognition is therefore that of how designers *formulate problems*. The second main area will be how designers *generate solutions*, since that is the over-riding aim and purpose of design activity: to generate a satisfactory design proposal. And the third main area will be the *process strategies* that

---

First presented at the international workshop on *Knowing and Learning in Design*, Georgia, Atlanta, USA, 1999, and first published as 'Design Cognition: Results from Protocol and Other Empirical Studies of Design Activity' in *Design Knowing and Learning: Cognition in Design Education*, edited by C M Eastman, W M McCracken and W C Newstetter, Elsevier, Oxford, UK, 2001.

designers employ, because there has been a lot of interest in design methodology – the understanding and structuring of design procedures – especially in the context of design education.

Problem Formulation

It is widely accepted that design 'problems' can only be regarded as a version of ill-defined problems. In a design project it is often not at all clear what 'the problem' is; it may have been only loosely defined by the client, many constraints and criteria may be un-defined, and everyone involved in the project may know that goals may be re-defined during the project. In design, 'problems' are often defined only in relation to ideas for their 'solution', and designers do not typically proceed by first attempting to define their problems rigorously.

One of the concerns in some other areas of design research has been to formulate design problems in *well*-defined ways. This is intended to overcome some of the inherent difficulties of attempting to solve ill-defined problems. However, designers' cognitive strategies are presumably based upon their normal need to resolve ill-defined problems. Thomas and Carroll (1979) carried out several observational and protocol studies of a variety of creative problem-solving tasks, including design tasks. One of their findings was that designers' behaviour was characterised by their treating the given problems *as though* they were ill-defined problems, for example by changing the goals and constraints, even when they could have been treated as well-defined problems. Thomas and Carroll concluded that: 'Design is a type of problem solving in which the problem solver views the problem or acts as though there is some ill-definedness in the goals, initial conditions or allowable transformations.' The implication is that designers will be designers, even when they could be problem-solvers.

Goal Analysis

This 'ill behaved' aspect of design behaviour has been noted even from the very earliest formal studies. Eastman (1970), in the earliest recorded design protocol study (of architectural design), found that: 'One approach to the problem was consistently expressed in all protocols. Instead of generating abstract relationships and attributes, then deriving the appropriate object to be considered, the subjects always generated a design element and then determined its qualities.' That is to say, the designer-subjects jumped to ideas for solutions (or partial solutions) before they had fully formulated the problem. This is a reflection of the fact that designers are solution-led, not problem-led; for designers, it is the evaluation of the solution that is important, not the analysis of the problem.

It is not just that problem-analysis is weak in design; even when problem goals and constraints are known or defined, they are not sacrosanct, and designers exercise the freedom to change goals and constraints, as understanding of the problem develops and definition of the solution proceeds. This was a feature of designer behaviour noted by Akin (1979) from his protocol studies of architects: 'One of the unique aspects of design behaviour is the constant generation of new task goals and redefinition of task constraints.' As Ullman *et al.* (1988) pointed out, only some constraints are 'given' in a design problem; other constraints are 'introduced' by the designer from domain knowledge, and others are 'derived' by the designer during the exploration of particular solution concepts.

The formulation of appropriate and relevant problem structures from the ill-defined problem of a design brief is not easy – it requires sophisticated skills in gathering and structuring information, and judging the moment to move on to solution generation. Christiaans and Dorst (1992), from protocol studies of junior and senior industrial design students, found that some students became stuck on information gathering, rather than progressing to solution generation. Interestingly, they found that this was not such a significant difficulty for junior students, who did not gather a lot of information, and tended to 'solve a simple problem', being unaware of a lot of potential criteria and difficulties. But they found that senior students could be divided into two types. The more successful group, in terms of the creativity quality of their solutions, 'asks less information, processes it instantly, and gives the impression of consciously building up an image of the problem. They look for and make priorities early on in the process.' The other group gathered lots of information, but for them 'gathering data was sometimes just a substitute activity for actually doing any design work' (Cross *et al.*, 1994).

A similar finding was reported by Atman *et al.* (1999), who found from their protocol analysis studies of engineering students that, for novices (freshmen with no design experience), '. . those subjects who spent a large proportion of their time defining the problem did not produce quality designs.' However, with senior students, Atman *et al.* did find that attention to 'problem scoping' (*i.e.*, 'adequately setting up the problem before analysis begins', including gathering a larger amount and wider range of problem-related information) did result in better designs. As with the industrial design students, some of the freshmen engineering students, it seemed, simply became stuck in problem-definition and did not progress satisfactorily into further stages of the design process.

Solution Focusing

Many studies suggest that designers move rapidly to early solution conjectures, and use these conjectures as means of exploring and defining problem-and-solution together. This is not a strategy employed by all problem-solvers, many of whom attempt to define or understand the problem fully before making solution attempts. This difference was observed by Lawson (1979), in his experiments on problem-solving behaviour in which he compared scientists with architects. Designers are solution-focused, unlike problem-focused scientists.

Lloyd and Scott (1994), from protocol studies of experienced engineering designers, found that a solution-focused approach appeared to be related to the degree and type of previous experience of the designers. They found that more experienced designers used more 'generative' reasoning, in contrast to the deductive reasoning employed more by less-experienced designers. In particular, designers with specific experience of the problem type tended to approach the design task through solution conjectures, rather than through problem analysis. Lloyd and Scott concluded that 'It is the variable of specific experience of the problem type that enables designers to adopt a conjectural approach to designing, that of framing or perceiving design problems in terms of relevant solutions.'

Co-evolution of Problem and Solution

Designers tend to use solution conjectures as the means of developing their understanding of the problem. Since 'the problem' cannot be fully understood in isolation from consideration of 'the solution', it is natural that solution conjectures should be used as a means of helping to explore and understand the problem formulation. As Kolodner and Wills (1996) observed, from a study of senior student engineering designers: 'Proposed solutions often directly remind designers of issues to consider. The problem and solution co-evolve.'

This interpretation of design as a co-evolution of solution *and* problem spaces has also been proposed by others, and has been found by Cross and Dorst (1998) in protocol studies of experienced industrial designers. They reported that: 'The designers start by exploring the [problem space], and find, discover, or recognise a partial structure. That partial structure is then used to provide them also with a partial structuring of the [solution space]. They consider the implications of the partial structure within the [solution space], use it to generate some initial ideas for the form of a design concept, and so extend and develop the partial structuring... They transfer the developed partial structure back into the [problem space], and again consider implications and extend the structuring of the [problem space]. Their goal ... is to create a matching problem-solution pair.'

Problem Framing

Designers are not limited to 'given' problems, but find and formulate problems within the broad context of the design brief. This is the characteristic of reflective practice identified by Schön (1983) as problem setting: 'Problem setting is the process in which, interactively, we *name* the things to which we will attend and *frame* the context in which we will attend to them.' This seems to characterise well what has been observed of the problem formulation aspects of design behaviour. Designers select features of the problem space to which they choose to attend (naming) and identify areas of the solution space in which they choose to explore (framing). Schön (1988) suggests that: 'In order to formulate a design problem to be solved, the designer must *frame* a problematic design situation: set its boundaries, select particular things and relations for attention, and impose on the situation a coherence that guides subsequent moves.'

This kind of problem framing has been noted often in studies of architects. Lloyd and Scott (1995), from their studies of (mostly senior-student) architects, reported that 'In each protocol there comes a time when the designer makes a statement that summarises how he or she *sees* the problem or, to be more specific, the structure of the situation that the problem presents.' They referred to this 'way of seeing the design situation' as the designer's 'problem paradigm'. As with their earlier studies of engineers, Lloyd and Scott found that the architects who had specific prior experience of the problem type had different approaches from their less-experienced colleagues: the experienced architects' approaches were characterised by strong problem paradigms, or 'guiding themes'. Cross and Clayburn Cross (1998) have also identified, from interviews and protocol studies, the importance of problem framing, or the use of a strong guiding theme or principle, in the design behaviour of outstanding, expert engineering designers. Darke (1979) also reported from interviews with outstanding architects that they used strong guiding themes as 'primary generators' for setting problem boundaries and solution goals.

Schön (1988) pointed out that 'the work of framing is seldom done in one burst at the beginning of a design process.' This was confirmed in Goel and Pirolli's (1992) protocol studies of several types of designers (architects, engineers and instructional designers). They found that 'problem structuring' activities not only dominated at the beginning of the design task, but also re-occurred periodically throughout the task.

Valkenburg and Dorst (1998) have attempted to develop and apply Schön's theory of reflective practice into team design activity, through a study of student industrial designers. In comparing a successful and an unsuccessful design team, Valkenburg and Dorst stressed the importance of the teams' problem framing. They identified five different frames used sequentially by the successful team during the project, in contrast to the single frame used by the unsuccessful team. The unsuccessful team also spent much greater amounts of time on 'naming' activities – *i.e.* on identifying potential problem features, rather than on developing solution concepts.

Solution Generation

The solution-focused nature of designer behaviour appears to be appropriate behaviour for responding to ill-defined problems. Such problems can perhaps never be converted to well-defined problems, and so designers quite reasonably adopt the more realistic strategy of finding a satisfactory solution, rather than expecting to be able to generate an optimum solution to a well-defined problem. However, this solution-focused behaviour also seems to have potential drawbacks. One such drawback might be the 'fixation' effect induced by existing solutions.

Fixation

A 'fixation' effect in design was suggested by Jansson and Smith (1991), who studied senior student and experienced professional mechanical engineers' solution responses to design problems. They compared groups of participants who were given a simple, written design brief, with those that were given the same brief but with the addition of an illustration of an existing solution to the set problem. They found that the latter groups appeared to be 'fixated' by the example design, producing solutions that contained many more features from the example design than did the solutions produced by the control groups. Jansson and Smith proposed that such fixation could hinder conceptual design if it prevents the designer from considering all of the relevant knowledge and experience that should be brought to bear on a problem. Designers may be too ready to re-use features of known existing designs, rather than to explore the problem and generate new design features.

Purcell and Gero (1991, 1993, 1996) undertook a series of experiments to verify and extend Jansson and Smith's findings on fixation. They studied and compared senior students in mechanical engineering and in industrial design. Early results suggested that mechanical engineers appeared to be much more susceptible to fixation than did industrial designers; the engineers' designs were substantially influenced by prior example designs, whereas the industrial designers appeared to be more fluent in producing a greater variety of designs, uninfluenced by examples. Purcell and Gero suggested that this might be a feature of the different educational programmes of engineers and designers, with the latter being more encouraged to generate diverse design solutions. In a further development of the study, however, Purcell and Gero explored engineers' and designers' responses when the example design was an innovative rather than a routine prior solution. Here they found that engineers became fixated in the traditional sense when shown a routine solution, *i.e.* incorporating features of the routine solution in their own solutions, but became fixated on the underlying *principle* of the innovative solution, *i.e.* producing new, innovative designs embodying the same principle. The industrial designers, however, responded in similar ways under both conditions, generating wide varieties of designs that were not substantially influenced by any of the prior designs. Purcell and Gero therefore concluded that the industrial designers seem to be 'fixated on being different', and that 'fixation' in design may exist in a number of forms.

It is not clear that 'fixation' is necessarily a bad thing in design. As mentioned above, Cross and Clayburn Cross (1998) have reported that outstanding expert designers exhibit a form of 'fixation' on their problem frame, or on a guiding theme or principle. Having established the 'frame' for a particular problem, these designers can be tenacious in their pursuit of solution concepts that fit the frame. Similar observations have been reported also by Candy and Edmonds (1996), in their study of an outstanding bicycle designer, and by Lawson (1994) in his studies of outstanding architects. This tenacious fixation seems to be found often amongst highly creative individuals.

Attachment to Concepts

Another form of 'fixation' that has been found to exist amongst designers is their attachment to early solution ideas and concepts. Although designers change goals and constraints as they design, they appear to hang on to their principal solution concept for as long as possible, even when detailed development of the scheme throws up unexpected difficulties and shortcomings in the solution concept. Some of the changing of goals and constraints during designing is associated with resolving such difficulties without having to start again with a major new concept. For example, from case studies of professional architectural design, Rowe (1987) observed that: 'A dominant influence is exerted by initial design ideas on subsequent problem-solving directions... Even when severe problems are encountered, a considerable effort is made to make the initial idea work, rather than to stand back and adopt a fresh point of departure.'

The same phenomenon was observed by Ullman *et al.* (1988), in protocol studies of experienced mechanical engineering designers. They found that 'designers typically pursue only a single design proposal,' and that 'there were many cases where major problems had been identified in a proposal and yet the designer preferred to apply patches rather than to reject the proposal outright and develop a better one.' A similar observation was also made by Ball *et al.* (1994), from their studies of senior students conducting 'real-world', final-year design projects in electronic engineering: 'When the designers were seen to generate a solution which soon proved less than satisfactory, they actually seemed loath to discard the solution and spend time and effort in the search for a better alternative. Indeed the subjects appeared to adhere religiously to their unsatisfactory solutions and tended to develop them laboriously by the production of various slightly improved versions until something workable was attained.'

Ball *et al.* regarded this behaviour as indicating a 'fixation' on initial concepts, and a reliance on a simple 'satisficing' design strategy in contrast to any more 'well-motivated' process of optimisation. They found it difficult to account for this apparently unprincipled design behaviour. Nevertheless, adherence to initial concepts and a satisficing strategy seem to be normal design behaviour. Guindon (1990b), in a study of experienced software designers, found that 'designers adopted a kernel solution very early in the session and did not elaborate any alternative solutions in depth. If designers retrieved alternative solutions for a sub-problem, they quickly rejected all but one alternative by a trade-off analysis using a preferred evaluation criterion.' In a very early study of design teams engaged on R&D projects for the space industry, Allen (1966) also found that preferred technical variants tended to become dominant early in a project, and that 'Once a technical approach becomes preferred over any other, it is not easily rejected. Furthermore, the longer it is in a dominant position, the more difficult it becomes to reject.' But this is not necessarily inappropriate design behaviour, because Allen found that the generation of alternative approaches during a project tended to be associated with teams producing poorer designs: 'Groups producing higher rated solutions generated fewer new approaches during the course of the project. There is some indication that these arise when the favoured approach encounters difficulty, and may sometimes be symptomatic of poorer performance on the part of the design team.'

However, in contrast to the 'fixation' findings reported above, in a study of senior industrial engineering students, Smith and Tjandra (1998) found that the quality of design solutions produced did appear to be dependent upon a willingness to reconsider early concepts. They experimented with nine groups of four students, undertaking an artificial design exercise based upon two-dimensional configurations of coloured triangles supposed to have different functional properties. Each member of a design team in the exercise played a different role (architect, thermal engineer, structural engineer, and cost estimator). One of Smith and Tjandra's findings was that 'The top three designs ... were the three groups that chose to scrap their initial design and to start afresh with a new design concept.' The successful players of this particular design game therefore seemed to be ones who were able and willing to overcome the possible fixation on an early concept. Perhaps it is worth emphasising that this study was based on a role-playing game and an artificial 'design' problem far removed from real-world design projects.

Generation of Alternatives

It may be that good designers produce good early concepts that do not need to be altered radically during further development. Or that good designers are able to modify their concepts rather fluently and easily as difficulties are encountered during development, without recourse to exploration of alternative concepts. Either way, it seems that designers are reluctant to abandon early concepts, and to generate ranges of alternatives. This does seem to be in conflict with a more 'principled' approach to design, as recommended by design theorists, and even to conflict with the idea that it is the exploration of solution concepts that assists the designer's problem understanding. Having more than one solution concept in play should promote a more comprehensive assessment and understanding of the problem.

Fricke (1993, 1996), from protocol studies of engineering designers, found that both generating few alternative concepts and generating a large number of alternatives were equally weak strategies, leading to poor design solutions. Where there was 'unreasonable restriction' of the search space (when only one or a very few alternative concepts were generated), designers became 'fixated' on concrete solutions too early. In the case of 'excessive expansion' of the search space (generating large numbers of alternative solution concepts), designers were then forced to spend time on organising and managing the set of variants, rather than on careful evaluation and modification of the alternatives. Fricke identified successful designers to be those operating a 'balanced search' for solution alternatives

Fricke also found that the degree of precision in the problem as it was presented to the designers influenced the generation of alternative solution concepts. When the problem was precisely specified, designers generated more solution variants; whereas with an imprecise assignment (for the same design task), designers tended to generate few alternative solution concepts. This perhaps indicates that the more active problem-framing required for an imprecise assignment leads more readily to preferred solution concepts. Designers given precise assignments have less scope for problem-framing, and generate a wider range of solution concepts in order to find a preferred concept.

Creativity

Designers themselves often emphasise the role of 'intuition' in the generation of solutions, and 'creativity' is widely regarded as an essential element in design thinking. Creative design is often characterised by the occurrence of a significant event, usually called the 'creative leap'. Recent studies of creative events in design have begun to shed more light on this previously mysterious (and often mystified) aspect of design.

Akin and Akin (1996) studied creative problem-solving behaviour first on a classic problem where a form of 'fixation' normally prevents people from finding a solution to the problem: the 'nine-dots' problem. (In this problem, nine dots are arranged in a 3 x 3 square, and subjects are invited to join all nine dots by drawing just four straight lines without lifting pen from paper. Subjects normally assume that they have to draw within the implicit outline of the square, whereas the solution requires extending the lines to new vertices outside of the square.) They then extended their study from the nine-dot problem into a study of a simple architectural design problem, and compared the protocols of a non-architect and an experienced architect in tackling this problem. In these studies, Akin and Akin were looking for cases of the 'sudden mental insight' (SMI) that is commonly reported in cases of creative problem solving. They referred to the 'fixation' effect, such as the implicit nine-dot square, as a 'frame of reference' (FR) that has to be broken out of in order to generate creative alternatives. They suggested that a SMI occurs when a subject perceives their own fixation within a standard FR, and simultaneously perceives a new FR. The new FR also has to include procedures for generating a solution to the problem. The experienced architect had such procedural knowledge, whereas the novice did not, and was not able to generate anything other than a very conventional solution. Akin and Akin conclude: 'Realising a creative solution, by breaking out of a FR, depends on simultaneously specifying a new set of FRs that restructure the problem in such a way that the creative process is enhanced. The new FRs must, at a minimum, specify an appropriate *representational* medium (permitting the explorations needed to go beyond those of the earlier FRs), a design *goal* (one that goes beyond those achievable within the earlier FRs), and a set of *procedures* consistent with the representation domain and the goals.'

This seems to be similar to Schön's concept of a 'frame' which permits and encourages the designer to explore new design 'moves' and to reflect on the discoveries arising from those moves. But 'frames' can clearly be negative conceptual structures, when they are inappropriate 'fixations', as well as positive, creative structures.

Akin and Akin's conclusions also resonate with the study by Cross (Chapter 4 in this volume) of the 'creative event' that occurred in a protocol study of teamwork in industrial design. The 'little vacuum-formed tray' concept appears to be the equivalent of a 'sudden mental insight', offering a new, creative 'frame of reference' meeting Akin and Akin's criteria, above.

It may be also that 'creative leaps' or 'sudden mental insights' are not so personal and idiosyncratic as has been promoted before. In protocol studies of experienced industrial designers, Cross and Dorst (1998) observed that all nine subjects reported *the same* 'creative breakthrough'. All nine linked together the same pieces of available information and used this as a basis for their solution concept. All nine appeared to think that this was a unique personal insight.

Sketching

Several researchers have investigated the ways in which sketching helps to promote creativity in design thinking. Sketching helps the designer to find unintended consequences, the surprises that keep the design exploration going in what Schön and Wiggins (1992) called the 'reflective conversation with the situation' that is characteristic of design thinking. Goldschmidt (1991) called it the 'dialectics of sketching': a dialogue between 'seeing that' and 'seeing as', where 'seeing that' is reflective criticism and 'seeing as' is the analogical reasoning and reinterpretation of the sketch that provokes creativity. Goel (1995) suggested that sketches help the designer to make not only 'vertical transformations' in the sequential development of a design concept, but also 'lateral transformations' within the solution space: the creative shift to new alternatives. Goel referred especially to the ambiguity inherent in sketches, and identified this as a positive feature of the sketch as a design tool.

It is not just formal or shape aspects of the design concept that are compiled by sketching; they also help the designer to identify and consider functional and other aspects of the design. Suwa, Purcell and Gero (1998) suggested that sketching serves at least three purposes: as an external memory device in which to leave ideas as visual tokens, as a source of visuo-spatial cues for the association of functional issues, and as a physical setting in which design thoughts are constructed in a type of situated action. Although the above studies refer mostly to sketching in architectural design, Ullman *et al.* (1990) also studied and emphasised the importance of sketching in mechanical engineering design, as have Kavakli *et al.* (1998) and McGown *et al.* (1998) in respect of product design. Verstijnen *et al.* (1998) studied differences between skilled sketchers (industrial design students) and unskilled sketchers, and concluded that it was the skilled sketchers who benefited from the externalisation of mental imagery.

Process Strategy

An aspect of concern in design methodology and related areas of design research has been the many attempts at proposing systematic models of the design process, and suggestions for methodologies or structured approaches that should lead designers efficiently towards a good solution. However, most design in practice still appears to proceed in a rather *ad-hoc* and unsystematic way. Many designers remain wary of systematic procedures that, in general, still have to prove their value in design practice.

Structured Processes

It is not clear whether learning a systematic process actually helps student designers. One study that has suggested that a systematic approach might be helpful to students was that of Radcliffe and Lee (1989). They studied fourteen senior students of mechanical engineering, working in small-groups (2-4) on a design project. In analysing the results, Radcliffe and Lee computed linear regression analyses of the subjects' design process sequence in comparison to an idealised, structured process of seven stages. They found that more 'efficient' processes (following closer to the supposed 'ideal') correlated positively with both quantity and quality of the subjects' design output: 'There was a positive correlation between the quality or effectiveness of a design and the degree to which the student follows a logical sequence of design processes.'

Fricke (1993, 1996) also studied a number of mechanical engineers, of varying degrees of experience and with varying exposures to education in systematic design processes. He found that designers following a 'flexible-methodical procedure' tended to produce good solutions. These designers worked reasonably efficiently and followed a fairly logical procedure, whether or not they had been educated in a systematic approach. In comparison, designers with too-rigid adherence to a methodical procedure (behaving 'unreasonably methodical'), or with very un-systematic approaches, produced mediocre or poor design solutions. It seems that, with or without an education in systematic design, designers need to exercise sophisticated strategic skills.

The occurrence of some relatively simple patterns of design process activity has often been suggested from anecdotal knowledge. For example, there has been a broad assumption that designing proceeds in cycles of analysis-synthesis-evaluation activities. Although such patterns of design process activity frequently have been proposed or hypothesised, there has been little empirical confirmation.

McNeill *et al.* (1998) were able to confirm some of these basic patterns in a study of electronics engineers, using subjects with varying degrees of experience, from senior students to very experienced professionals. They were able to confirm that, 'In addition to the short-term cycles [of analysis-synthesis-evaluation], there is a trend over the whole design episode to begin by spending most of the time analysing the problem, then mainly synthesising the solution and finishing by spending most time on the evaluation of the solution.' They also confirmed a supposed progression through the design process from first considering required functions, then structure of potential solutions, and then the behaviour of those solutions. Their general, if unsurprising conclusion was that: 'A designer begins a conceptual design session by analysing the functional aspects of the problem. As the session progresses, the designer focuses on the three aspects of function, behaviour and structure, and engages in a cycle of analysis, synthesis and evaluation. Towards the end of the design session, the designer's activity is focused on synthesising structure and evaluating the structure's behaviour.'

Opportunism

In contrast to studies that confirm the prevalence and relevance of fairly structured design behaviour, there have also been reports of some studies that emphasised the 'opportunistic' behaviour of designers. This emphasis has been on designers' deviations from a structured plan or methodical process into the 'opportunistic' pursuit of issues or partial solutions that catch the designer's attention. For example, Visser (1990) made a longitudinal study of an experienced mechanical engineer, preparing a design specification. The engineer claimed to be following a structured approach, but Visser found frequent deviations from this plan. 'The engineer had a hierarchically structured plan for his activity, but he used it in an opportunistic way. He used it only as long as it was profitable from the point of view of cognitive cost. If more economical cognitive actions arose, he abandoned it.' Thus Visser regarded reducing 'cognitive cost' – *i.e.* the cognitive load of maintaining a principled, structured approach – as a major reason for abandoning planned actions and instead delving into, for example, confirming a partial solution at a relatively early stage of the process.

From protocol studies of three experienced software system designers, Guindon (1990a) also emphasised the 'opportunistic' nature of design activities. Guindon stressed that 'designers frequently deviate from a top-down approach. These results cannot be accounted for by a model of the design process where problem specification and understanding precedes solution development and where the design solution is elaborated at successively greater levels of detail in a top-down manner.' Guindon observed the interleaving of problem specification with solution development, 'drifting' through partial solution development, and jumps into exploring suddenly-recognised partial solutions, which she categorised as major causes of 'opportunistic solution development'. She also referred to 'cognitive cost' as one possible explanation for such behaviour: 'Designers find it advantageous to follow a train of thought temporarily, thus arriving at partial solutions at little cognitive cost.'

Ball and Ormerod (1995) criticised a too-eager willingness to emphasise 'opportunism' in design activity. In their studies of expert electronics engineers they found very few deviations from a top-down, breadth-first design strategy. But they did find some significant deviations occurring, when designers made a rapid depth-first exploration of a solution concept in order to assess its viability. Ball and Ormerod did not regard such occasional depth-first explorations as implying the abandonment of a structured approach. Instead, they suggested that expert designers will normally use a mixture of breadth-first and depth-first approaches: 'Much of what has been described as opportunistic behaviour sits comfortably within a structured top-down design framework in which designers alternate between breadth-first and depth-first modes.' Ball and Ormerod were concerned that 'opportunism' seemed to imply unprincipled design behaviour, 'a non-systematic and heterarchical process' in contrast to the assumed ideal of a systematic and hierarchical process. However, rather than regarding opportunism as unprincipled design behaviour, Guindon had suggested it might be inevitable in design: 'These deviations are not special cases due to bad design habits or performance breakdowns but are, rather, a natural consequence of the ill-structuredness of problems in the early stages of design.' So it may be that we should not equate 'opportunistic' with 'unprincipled' behaviour in design, but rather that we should regard 'opportunism' as characteristic of expert design behaviour.

Modal Shifts

An aspect of cognitive strategy that emerges from several studies is that, especially during creative periods of conceptual design, designers alternate rapidly in shifts of attention between different aspects of their task, or between different modes of activity. Akin and Lin (1996), in their protocol study of an experienced engineering designer, first identified the occurrence of 'novel design decisions' (NDDs). These, in contrast to routine design decisions, are decisions that are critical to the development of the design concept. Akin and Lin also segmented the designer's activities into three modes, drawing, examining and thinking. Then, allowing for some implicit overlap or carry-over of the designer's attention from one segment to another, they represented the designer's activities in terms of single-, dual- or triple-mode periods. They found a significant correlation between the triple-mode periods and the occurrence of the NDDs: 'Six out of a total of eight times a novel design decision was made, we found the subject alternating between these three activity modes (examining-drawing-thinking) in rapid succession.' Akin and Lin are cautious about drawing any inference of causality, concluding only that 'Our data suggest that designers explore their domain of ideas in a variety of activity modes ... when they go beyond routine decisions and achieve design breakthroughs.'

Some studies of student designers have also noted the apparent importance of frequent shifts of attention or activity mode in influencing either the creativity or overall quality of the design concepts produced. For example, in their protocol studies of junior and senior students of industrial design, Cross *et al.* (1994) segmented the students' activities into the three modes of gathering information, sketching and reflecting. They suggested that the more successful students (in producing creative design concepts) were those who showed evidence of rapid alternation between the activity modes. Also, Atman *et al.* (1999), from their study of freshmen and senior engineering design students, suggested that overall quality of design concepts was related to rapid alternation of activities, which they measured as transitions between design steps such as gathering information, generating ideas and modelling.

Novices and Experts

Novice behaviour is usually associated with a 'depth-first' approach to problem solving, *i.e.* sequentially identifying and exploring sub-solutions in depth, whereas the strategies of experts are usually regarded as being predominantly top-down and breadth-first approaches. But this may be too simplistic a view of the reality of process strategy in design. Ball and Ormerod's (1995) comments about top-down, structured approaches *versus* 'opportunism' have been noted above. They concluded that 'it would be surprising if it is practicable for expert designers to adopt a purely breadth-first or depth-first approach. Indeed, a flexible mixture of modes is a more psychologically realistic control structure for expert design.' They suggested that, whilst a depth-first approach minimises cognitive load, a breadth-first approach minimises commitment and optimises design time and effort. Those suggestions would also quite reasonably reflect the respective concerns and strategies that we might expect of novices and experts.

Many of the classic studies of expertise have been based on examples of game-playing (*e.g.* chess), or on comparisons of experts *versus* novices in solving routine problems (*e.g.* physics). These are all well-defined problems, whereas designers characteristically deal with ill-defined problems. Some studies of expertise in fields such as creative writing and computer programming (Holyoak, 1991; Adelson and Solway, 1988) where problems are more ill-defined, do suggest some parallels with observations of expert designers. These studies suggest that some of the 'standard' results from studies of expertise do not match with results from studies of expertise in creative domains. For example, creative experts will define the given task so that it is problematic – *i.e.* deliberately treat it as ill-defined – which is contrary to the assumption that experts will generally solve a problem in the 'easiest' way, or certainly with more ease than novices. In some ways, therefore, creative experts treat problems as 'harder' problems than novices do. Creative experts are also reported as solving similar tasks from first principles each time, rather than recalling previous solutions.

Göker (1997) compared novices and experts performing design-related problem solving tasks – the computer-simulated construction of 'machines' from catalogues of parts, to achieve certain objectives. A particularly unusual aspect of this study was the experimental method, based on the use of electro-encephalograph (EEG) records. Göker found that experts (subjects skilled in the use of the computer simulation) used more of the visuo-spatial regions of their brains than did the novices, who used more of their brain regions associated with verbal-abstract reasoning. The implication is that experts do not 'reason' towards a design concept in an abstract way, but rely more on their experience and on visual information.

Issues in Design Cognition

In this chapter I have surveyed a wide range of empirical studies of design cognition, and attempted to draw out some of the issues that have emerged in such studies. I have taken a cross-disciplinary view, and looked for comparisons across the different domains of professional design practice. There has been a number of striking similarities identified in design activity, independent of professional domain, suggesting that design cognition is indeed a domain-independent phenomenon.

I have concentrated on protocol and similar formalised methods of study, and I have therefore omitted a wide range of other kinds of studies that also have relevant and important contributions to make to the understanding of design cognition and the nature of design activity. Protocol analysis has some severe limitations as a research method for investigating design activity. For instance, it is extremely weak in capturing non-verbal thought processes, which are so important in design work (Lloyd *et al.*, 1995). Dorst and Cross (1997) concluded from the Delft Design Protocols Workshop that protocol analysis provides a very valuable but highly specific research technique, capturing a few aspects of design thinking in detail, but failing to encompass many of the broader realities of design in context. Other kinds of study, which attempt to capture a broader view, include detailed   observation of industrial practice, such as Frankenberger and Badke-Schaub (1998), and ethnographic methods, such as Bucciarelli (1994). There has also been valuable historical work, such as Ferguson's (1992) study of the role of drawing in engineering, and significant theoretical contributions to identifying fundamental aspects of design reasoning and logic, such as Roozenburg (1993).

The range and number of studies surveyed in this chapter suggest that the field of empirical studies of design activity is continuing to grow, and a number of shared issues has been identified. In many cases, these issues remain unresolved, and there is therefore still considerable work to be done to establish a robust and reliable understanding of design cognition.

Summary: Problem Formulation

Goal Analysis
Designers appear to be 'ill-behaved' problem solvers, in that they do not spend much time and attention on defining the problem. However, this seems to be appropriate behaviour, since some studies have suggested that over-concentration on problem definition does not lead to successful design outcomes. It appears that successful design behaviour is based not on extensive problem analysis, but on adequate 'problem scoping' and on a focused or directed approach to gathering problem information and prioritising criteria. Setting and changing goals are inherent elements of design activity.

Solution Focusing
Designers are solution-focused, not problem-focused. This appears to be a feature of design cognition which comes with education and experience in designing. In particular, experience in a specific problem domain enables designers to move quickly to identifying a problem 'frame' and proposing a solution conjecture.

Co-evolution of Problem and Solution
The concept of 'co-evolution' of both the problem and its solution has been proposed to describe how designers develop both aspects together in conceptual stages of the design process. The designer's attention oscillates between the two, forming partial structurings of the two 'spaces' of problem and solution. Designing appears to be an 'appositional' search for a matching problem-solution pair, rather than a propositional argument from problem to solution.

Problem Framing
Processes of structuring and formulating the problem are frequently identified as key features of design activity. The concept of 'problem framing' perhaps seems to capture best the nature of this activity. Successful, experienced and – especially – outstanding designers are repeatedly found in various studies to be pro-active in problem framing, actively imposing their view of the problem and directing the search for solution conjectures.

Summary: Solution Generation

Fixation
'Fixation' seems to be a double-edged feature of design activity, in that it can lead to conservative, routine design or – perhaps only when exercised by outstanding designers – to creative, innovative design. There may be differences between educational programmes of engineers and industrial designers (and probably architects) which lead engineers more readily to fixate on features of prior design solutions.

Attachment to Concepts

Designers become readily attached to single, early solution concepts and are reluctant to abandon them in the face of difficulties in developing these concepts into satisfactory solutions. This seems to be a weak feature of design behaviour, which may be susceptible to change through education. However, trying to change the 'unprincipled' and 'ill-behaved' nature of conventional design activity may be working against aspects that are actually effective and productive features of intuitive design cognition.

Generation of Alternatives

Generating a wide range of alternative solution concepts is another aspect of design behaviour which is recommended by theorists and educationists but appears not to be normal design practice. Generating a very wide range of alternatives may not be a good thing: some studies have suggested that a relatively limited amount of generation of alternatives may be the most appropriate strategy.

Creativity

Creative thinking has tended to be regarded as mysterious, but new explanatory descriptions of creativity in design are beginning to emerge from empirical studies. In particular, it no longer seems correct to promote the key feature of creative design as dependent upon an intuitive, heroic 'creative leap' from problem to solution. Problem framing, co-evolution, and conceptual bridging between problem space and solution space seem to be better descriptors of what actually happens in creative design.

Sketching

The key 'tool' to assist design cognition remains the traditional sketch. It seems to support and facilitate the uncertain, ambiguous and exploratory nature of conceptual design activity. Sketching is tied-in very closely with features of design cognition such as the generation and exploration of tentative solution concepts, the identification of what needs to be known about the developing concept, and especially the recognition of emergent features and properties. Studies of the role of sketching have all emphasised its inherent power as a design aid.

Summary: Process Strategy

Structured Process

Following a reasonably-structured process seems to lead to greater design success. However, rigid, over-structured approaches do not appear to be successful. The key seems to be flexibility of approach, which comes from a rather sophisticated understanding of process strategy and its control.

Opportunism

'Opportunistic' behaviour sounds like another feature of the characteristically 'unprincipled', 'ill-behaved' activity of designers. As with some other aspects of intuitive design behaviour, it may be that we should not equate 'opportunistic' with 'unprincipled' behaviour in design, but rather that we should regard 'opportunism' as characteristic of expert design behaviour. The 'cognitive cost' of apparently more principled, structured behaviour may actually be higher than can be reasonably sustained, or can be justified by quality of outcome.

Modal Shifts

It has been noticed in some studies that creative, productive design behaviour seems to be associated with frequent switching of types of cognitive activity. There is no clear explanation for this observation, but it may be related to the need to make rapid explorations of problem and solution in tandem.

Novices and Experts

Conventional wisdom about the nature of problem-solving expertise seems often to be contradicted by the behaviour of expert designers. In design education we must therefore be very wary about importing models of behaviour from other fields. Empirical studies of design activity have frequently found 'intuitive' features of design behaviour to be the most effective and relevant to the intrinsic nature of design. Some aspects of design theory, however, have tried to develop counter-intuitive models and prescriptions for design behaviour. We still need a much better understanding of what constitutes expertise in design, and how we might assist novice students to gain that expertise.

7

Design as a Discipline

I would like to begin this chapter with a brief review of some of the historical concerns with establishing a relationship between design and science. These concerns emerged strongly at two important periods in the modern history of design: in the 1920s, with a search for scientific design products, and in the 1960s, with a search for scientific design process. The 40-year cycle in these concerns appears to be coming around again, and we might expect to see the re-emergence of design-science concerns in the 2000s.

A desire to 'scientise' design can be traced back to ideas in the 20th-Century modern movement in design. For example, in the early 1920s, the *De Stijl* protagonist, Theo van Doesburg (1923) expressed this perception of a new spirit in art and design: 'Our epoch is hostile to every subjective speculation in art, science, technology, *etc*. The new spirit, which already governs almost all modern life, is opposed to animal spontaneity, to nature's domination, to artistic flummery. In order to construct a new object we need a method, that is to say, an objective system.' A little later, the architect Le Corbusier (1929) wrote about the house as an objectively-designed 'machine for living': 'The use of the house consists of a regular sequence of definite functions. The regular sequence of these functions is a traffic phenomenon. To render that traffic exact, economical and rapid is the key effort of modern architectural science.' In both of these comments, and throughout much of the Modern Movement, we see a desire to produce works of art and design based on objectivity and rationality, that is, on the values of science.

These aspirations to scientise design surfaced strongly again in the 'design methods movement' of the 1960s. The Conference on Design Methods, held in London in September, 1962 (Jones and Thornley, 1963) is generally regarded as the event which marked the launch of design methodology as a subject or field of enquiry. The desire of the new movement was even more strongly than before to base design process (as well as the products of design) on objectivity and rationality. The origins of this emergence of new design methods in the 1960s lay

First presented as 'Designerly Ways of Knowing: Design Discipline *vs* Design Science' at the international conference *Design+Research*, Politecnico di Milano, Italy, 2000.

in the application of novel, scientific and computational methods to the novel and pressing problems of the 2nd World War – from which came civilian developments such as operations research and management decision-making techniques.

The 1960s was heralded as the 'design science decade' by the radical technologist Buckminster Fuller, who called for a 'design science revolution', based on science, technology and rationalism, to overcome the human and environmental problems that he believed could not be solved by politics and economics. From this perspective, the decade culminated with Herbert Simon's (1969) outline of 'the sciences of the artificial' and his specific plea for the development of 'a science of design' in the universities: 'a body of intellectually tough, analytic, partly formalizable, partly empirical, teachable doctrine about the design process.'

However, in the 1970s there emerged a backlash against design methodology and a rejection of its underlying values, notably by some of the early pioneers of the movement. Christopher Alexander, who had originated a rational method for architecture and planning (Alexander, 1964), now said: 'I've disassociated myself from the field... There is so little in what is called "design methods" that has anything useful to say about how to design buildings that I never even read the literature anymore... I would say forget it, forget the whole thing' (Alexander, 1971). Another leading pioneer, J. Christopher Jones (1977) said: 'In the 1970s I reacted against design methods. I dislike the machine language, the behaviourism, the continual attempt to fix the whole of life into a logical framework.'

To put the quotations of Alexander and Jones into context it may be necessary to recall the social/cultural climate of the late-1960s – the campus revolutions and radical political movements, the new liberal humanism and rejection of conservative values. But also it had to be acknowledged that there had been a lack of success in the application of 'scientific' methods to everyday design practice. Fundamental issues were also raised by Rittel and Webber (1973), who characterised design and planning problems as 'wicked' problems, fundamentally un-amenable to the techniques of science and engineering, which dealt with 'tame' problems.

Nevertheless, design methodology continued to develop strongly, especially in engineering and some branches of industrial design. (Although there may still have been very limited evidence of practical applications and results.) The fruits of this work emerged in a series of books on engineering design methods and methodology in the 1980s. Just to mention some English-language ones, these included Tjalve (1979), Hubka (1982), Pahl and Beitz (1984), French (1985), Cross (1989), Pugh (1991).

Another significant development throughout the 1980s and into the 1990s was the emergence of new journals of design research, theory and methodology. Just to refer, again, to English-language publications, these included *Design Studies* in 1979, *Design Issues* in 1984, *Research in Engineering Design* in 1989, the *Journal of Engineering Design* in 1990, *Languages of Design* in 1993 and the *Design Journal* in 1997.

Despite the apparent scientific basis (or bias) of much of their work, design methodologists also sought from the earliest days to make distinctions between design and science, as reflected in the following quotations.

Scientists try to identify the components of existing structures, designers try to shape the components of new structures. Alexander (1964)

The scientific method is a pattern of problem-solving behaviour employed in finding out the nature of what exists, whereas the design method is a pattern of behaviour employed in inventing things of value which do not yet exist. Science is analytic; design is constructive. Gregory (1966)

There may indeed be a critical distinction to be made: method may be vital to the practice of science (where it validates the results) but not to the practice of design (where results do not have to be repeatable, and in most cases must *not* be repeated, or copied). The Design Research Society's 1980 conference on 'Design:Science:Method' (Jacques and Powell, 1981) gave an opportunity to air many of these considerations. The general feeling from that conference was perhaps that it was time to move on from making simplistic comparisons and distinctions between science and design; that perhaps there was not so much for design to learn from science after all, and rather that perhaps science had something to learn from design. Cross *et al.* (1981) claimed that the epistemology of science was, in any case, in disarray, and therefore had little to offer an epistemology of design. Glynn (1985) later suggested that 'it is the epistemology of design that has inherited the task of developing the logic of creativity, hypothesis innovation or invention that has proved so elusive to the philosophers of science.'

Despite several attempts at clarification (see de Vries, Cross and Grant, 1993) there remains some confusion about the design-science relationship. Let us at least try to clarify three different interpretations of this concern with the relationship between science and design: (a) scientific design, (b) design science, and (c) a science of design.

Scientific Design

As I noted above, the origins of design methods lay in 'scientific' methods, similar to decision theory and the methods of operational research. The originators of the 'design methods movement' also realised that there had been a change from the craftwork of pre-industrial design to the mechanisation of industrial design – and perhaps some even foresaw the emergence of a post-industrial design. The reasons advanced for developing new methods were often based on the assumption that modern, industrial design had become too complex for intuitive methods.

The first half of the twentieth century had seen the rapid growth of scientific underpinnings in many types of design – *e.g.* materials science, engineering science, building science, behavioural science. One view of the design-science relationship is that, through this reliance of modern design upon scientific knowledge, through the application of scientific knowledge in practical tasks, design 'makes science visible' (Willem, 1990).

So we might agree that *scientific design* refers to modern, industrialised design – as distinct from pre-industrial, craft-oriented design – based on scientific knowledge but utilising a mix of both intuitive and non-intuitive design methods. 'Scientific design' is probably not a controversial concept, but merely a reflection of the reality of modern design practice.

Design Science

'Design Science' was a term perhaps first used by Buckminster Fuller, but it was adapted by Gregory (1966) into the context of the 1965 conference on 'The Design Method'. The concern to develop a design science thus led to attempts to formulate *the* design method – a single rationalised method, as 'the scientific method' was supposed to be. Others, too, have had the development of a 'design science' as their aim; for example, Hubka and Eder (1987), originators of the *Workshop Design Konstruction* (WDK) and a major series of international conferences on engineering design (ICED), also formed 'The International Society for Design Science'. Earlier, in Germany, Hansen (1974) had stated the aim of design science as being to 'recognize laws of design and its activities, and develop rules'. This would seem to be design science constituted simply as 'systematic design' – the procedures of designing organized in a systematic way. Hubka and Eder regarded this as a narrower interpretation of design science than their own, which was: 'Design science comprises a collection (a system) of logically connected knowledge in the area of design, and contains concepts of technical information and of design methodology... Design science addresses the problem of determining and categorizing all regular phenomena of the systems to be designed, and of the design process. Design science is also concerned with deriving from the applied knowledge of the natural sciences appropriate information in a form suitable for the designer's use.' This definition extends beyond 'scientific design', in including systematic knowledge of design process and methodology as well as the scientific/technological underpinnings of design of artefacts.

So we might conclude that *design science* refers to an explicitly organised, rational and wholly systematic approach to design; not just the utilisation of scientific knowledge of artefacts, but design in some sense a scientific activity itself. This is certainly a controversial concept, challenged by many designers and design theorists. As Grant (1979) wrote: 'Most opinion among design methodologists and among designers holds that the act of designing itself is not and will not ever be a scientific activity; that is, that designing is itself a non-scientific or a-scientific activity.'

Science of Design

However, Grant also made it clear that 'the study of designing may be a scientific activity; that is, design as an activity may be the subject of scientific investigation.' There remains some confusion between concepts of design science and of a science of design, since a 'science of design' seems to imply (or for some people has an aim of) the development of a 'design science'. But the concept of a science of design has been clearly stated by Gasparski and Strzalecki (1990): 'The science of design (should be) understood, just like the science of science, as a federation of subdisciplines having design as the subject of their cognitive interests'.

In this latter view, therefore, the science of design is the *study* of design – something similar to what I have elsewhere defined 'design methodology' to be; the study of the principles, practices and procedures of design. For me, design methodology 'includes the study of how designers work and think, the establishment of appropriate structures for the design process, the development and application of new design methods, techniques and procedures, and reflection on the nature and extent of design knowledge and its application to design problems' (Cross, 1984). The *study* of design leaves open the interpretation of the *nature* of design.

So let us agree here that the *science of design* refers to that body of work which attempts to improve our understanding of design through 'scientific' (*i.e.*, systematic, reliable) methods of investigation. And let us be clear that a 'science of design' is not the same as a 'design science'.

Design as a Discipline

Donald Schön (1983) explicitly challenged the positivist doctrine underlying much of the 'design science' movement, and offered instead a constructivist paradigm. He criticised Simon's 'science of design' for being based on approaches to solving well-formed problems, whereas professional practice throughout design and technology and elsewhere has to face and deal with 'messy, problematic situations'. Schön proposed instead to search for 'an epistemology of practice implicit in the artistic, intuitive processes which some practitioners do bring to situations of uncertainty, instability, uniqueness, and value conflict,' and which he characterised as 'reflective practice'. Schön appeared to be more prepared than his positivist predecessors to put trust in the abilities displayed by competent practitioners, and to try to explicate those competencies rather than to supplant them. This approach has been developed particularly in the series of workshops and conferences known as the 'Design Thinking Research Symposia', beginning in 1991 (Cross, *et al.*, 1992).

Despite the positivist, technical-rationality basis of *The Sciences of the Artificial,* Simon (1969) did propose that 'the science of design' could form a fundamental, common ground of intellectual endeavour and communication across the arts, sciences and technology. What he suggested was that the study of design could be an interdisciplinary study accessible to all those involved in the creative activity of making the artificial world (which effectively includes all mankind). For example, Simon wrote that 'Few engineers and composers … can carry on a mutually rewarding conversation about the content of each other's professional work. What I am suggesting is that they *can* carry on such a conversation about design, can begin to perceive the common creative activity in which they are both engaged, can begin to share their experiences of the creative, professional design process.'

This, it seems to me, is the challenge for a broad and catholic approach to design research – to construct a way of conversing about design that is at the same time both interdisciplinary and disciplined. We do not want conversations that fail to connect between sub-disciplines, that fail to reach common understanding, and that fail to create new knowledge and perceptions of design. It is the paradoxical task of creating an interdisciplinary discipline. Design as a discipline, rather than design as a science. This discipline seeks to develop domain-independent approaches to theory and research in design. The underlying axiom of this discipline is that there are forms of knowledge peculiar to the awareness and ability of a designer, independent of the different professional domains of design practice. Just as the other intellectual cultures in the sciences and the arts concentrate on the underlying forms of knowledge peculiar to the scientist or the artist, so we must concentrate on the 'designerly' ways of knowing, thinking and acting.

Many researchers in the design world have been realising that design practice does indeed have its own strong and appropriate intellectual culture, and that we must avoid swamping our design research with different cultures imported either from the sciences or the arts. This does not mean that we completely ignore these other cultures. On the contrary, they have much stronger histories of enquiry, scholarship and research than we have in design. We need to draw upon those histories and traditions where appropriate, whilst building our own intellectual culture, acceptable and defensible in the world on its own terms. We have to be able to demonstrate that standards of rigour in our intellectual culture at least match those of the others.

Design Research

At the 1980 'Design:Science:Method' conference of the Design Research Society, Archer (1981) gave a simple but useful definition of research, which is that 'Research is systematic enquiry, the goal of which is knowledge'. Our concern in design research has to be the development, articulation and communication of design knowledge. Where do we look for this knowledge? I believe that it has three sources: people, processes and products.

Design knowledge resides firstly in *people*: in designers especially, but also in everyone to some extent. Designing is a natural human ability. Other animals do not do it, and machines (so far) do not do it. We often overlook the fact that people are naturally very good at design. We should not underplay our abilities as designers: many of the most valued achievements of humankind are works of design, including anonymous, vernacular design as well as the 'high design' of professionals.

One immediate subject of design research, therefore, is the investigation of this human ability – of how people design. This suggests, for example, empirical studies of design behaviour, but it also includes theoretical deliberation and reflection on the nature of design ability. It also relates strongly to considerations of how people learn to design, to studies of the development of design ability in individuals and how that development might best be nurtured in design education.

Design knowledge resides secondly in *processes*: in the tactics and strategies of designing. A major area of design research is methodology: the study of the processes of design, and the development and application of techniques which aid the designer. Much of this research revolves around the study of modelling for design purposes. Modelling is the 'language' of design. Traditional models are the sketches and drawings of proposed design solutions, but which in contemporary terms now extend to 'virtual reality' models. The use of computer-based models has stimulated a wealth of research into design processes.

Thirdly, we must not forget that design knowledge resides in *products* themselves: in the forms and materials and finishes which embody design attributes. Much everyday design work entails the use of precedents or previous exemplars – not because of laziness by the designer but because the exemplars actually contain knowledge of what the product should be. This is certainly true in craft-based design: traditional crafts are based on the knowledge implicit within the object itself of how best to shape, make and use it. This is why craft-made products are usually copied very literally from one example to the next, from one generation to the next.

As with the design knowledge that resides in people, we would be foolish to disregard or overlook this informal product knowledge simply because it has not been made explicit yet – that is a task for design research. So too is the development of more formal knowledge of shape and configuration – theoretical studies of design morphology. These may be concerned as much with the semantics as with the syntax of form, or may be concerned with prosaic matters of efficiency and economy, or with relationships between form and context – whether ergonomics or environment.

My own taxonomy of the field of design research would therefore fall into three main categories, based on people, process and products:

- design epistemology – study of designerly ways of knowing
- design praxiology – study of the practices and processes of design
- design phenomenology – study of the form and configuration of artefacts

What has been happening in the field of design research is that there has been a growing awareness of the intrinsic strengths and appropriateness of design thinking within its own context, of the validity of a form of 'design intelligence'. There has been a growing acceptance of design on its own terms, a growing acknowledgement and articulation of design as a discipline. We have come to realise that we do not have to turn design into an imitation of science; neither do we have to treat design as a mysterious, ineffable art. We recognize that design has its own distinct intellectual culture.

But there is also some confusion and controversy over the nature of design research. I believe that examples of 'best practice' in design research have in common the following characteristics, which I have borrowed and adapted from lecture notes by Bruce Archer.

Good research is:

Purposive   - based on identification of an issue or problem worthy and capable of investigation

Inquisitive   - seeking to acquire new knowledge

Informed   - conducted from an awareness of previous, related research

Methodical   - planned and carried out in a disciplined manner

Communicable   - generating and reporting results which are testable and accessible by others

These characteristics are normal features of good research in any discipline. I do not think that such normal criteria inhibit or preclude research that is 'designerly' in its origins and intentions. However, they would exclude works of so-called research that fail to communicate, are undisciplined or ill-informed, and therefore add nothing to the body of knowledge of the discipline.

We also need to draw a distinction between works of practice and works of research. I do not see how normal works of practice can be regarded as works of research. The whole point of doing research is to extract reliable knowledge from either the natural or artificial world, and to make that knowledge available to others in re-usable form. This does not mean that works of design practice must be wholly excluded from design research, but it does mean that, to qualify as research, there must be reflection by the practitioner on the work, and the communication of some re-usable results from that reflection.

One of the dangers in this new field of design research is that researchers from other, non-design, disciplines will import methods and approaches that are inappropriate to developing the understanding of design. Researchers from psychology or computer science, for example, have tended to assume that there is 'nothing special' about design as an activity for investigation, that it is just another form of 'problem solving' or 'information processing'. However, developments in artificial intelligence and other computer modelling in design have perhaps served mainly to demonstrate just how high-level and complex is the cognitive ability of designers, and how much more research is needed to understand it. Better progress seems to be made by designer-researchers, and for this reason the recent growth of conferences, workshops and symposia, featuring a new generation of designer-researchers, is proving extremely useful in developing the methodology of design research. As design grows as a discipline with its own research base, so we can hope that there will be a growth in the number of emerging designer-researchers.

Another of the dangers is that researchers adhere to underlying paradigms of which they are only vaguely aware. We need to develop this intellectual awareness within our community. An example of developing this awareness is the work of Dorst (1997), in making an explicit analysis and comparison of the paradigms underlying the approach of Herbert Simon, on the one hand, and Donald Schön on the other. Simon's positivism leads to a view of design as 'rational problem solving', and Schön's constructivism leads to a view of design as 'reflective practice'. These two might appear to be in conflict, but Dorst's use of the two paradigms in analysing design activity leads him to the view that the different paradigms have complementary strengths for gaining an overview of the whole range of activities in design.

We are still building the appropriate paradigm for design research. I have made it clear that my personal 'touch-stone' theory for this paradigm is that there are 'designerly ways of knowing'. I believe that building such a paradigm will be helpful, in the long run, to design practice and design education, and to the broader development of the intellectual culture of our world of design.

References

Adelson, B and Solway, E (1985) The Role of Domain Experience in Software Design *IEEE Transactions on Software Engineering* Vol 11, pp. 1351–1360

Adelson, B and Solway, E (1988) A Model of Software Design, in M Chi, R Glaser and M Farr (eds.) *The Nature of Expertise*, Erlbaum, Hillsdale, NJ , USA

Akin, Ö (1979) An Exploration of the Design Process *Design Methods and Theories* Vol 13, No 3/4, pp. 115–119

Akin, Ö and Akin, C (1996) Frames of Reference in Architectural Design: analysing the hyper-acclamation (Aha!) *Design Studies* Vol 17, No 4, pp. 341–361

Akin, Ö and Lin, C (1996) Design Protocol Data and Novel Design Decisions, in Cross, N. *et al.* (eds.), *Analysing Design Activity*, John Wiley and Sons Ltd., Chichester, UK

Alexander, C (1964) *Notes on the Synthesis of Form* Harvard University Press, Cambridge, MA, USA

Alexander, C (1971) The State of the Art in Design Methods *DMG Newsletter*, Vol 5, No 3, pp. 3–7

Alexander, C *et al.* (1979) *A Pattern Language* Oxford University Press, New York, USA

Amabile, T (1982) Social Psychology of Creativity: a consensual assessment technique *Journal of Personality and Social Psychology* Vol. 43, pp. 997–1013

Archer, L B (1965) *Systematic Method for Designers*, The Design Council, London, UK. Reprinted in N Cross (ed.) (1984), *Developments in Design Methodology*, John Wiley and Sons Ltd., Chichester, UK

Archer, L B (1981) A View of the Nature of Design Research, in Jacques, R and Powell, J (eds.), *Design:Science:Method*, Westbury House, Guildford, UK

Ball, L J, Evans, J, *et al.* (1994) Cognitive Processes in Engineering Design: a longitudinal study *Ergonomics* Vol 37, No 11, pp. 1753–1786

Ball, L J and Ormerod, T (1995) Structured and Opportunistic Processing in Design: a critical discussion *International Journal of Human-Computer Studies* Vol 43, pp. 131–151

Blakeslee, T R (1980) *The Right Brain* Macmillan, London, UK

Boden, M (1990) *The Creative Mind: Myths and Mechanisms* Weidenfield and Nicolson, London, UK

Bogen, J E (1969) The Other Side of the Brain II: an appositional mind *Bulletin of the Los Angeles Neurological Societies* Vol 34, No 3

Bucciarelli, L (1994) *Designing Engineers* MIT Press, Cambridge, MA, USA

Candy, L and Edmonds, E (1996) Creative Design of the Lotus Bicycle *Design Studies* Vol 17, No 1, pp. 71–90

Casti, J (1998) *The Cambridge Quintet* Little, Brown, London, UK

Christiaans, H (1992) *Creativity in Design: The Role of Domain Knowledge in Designing*, Lemma, Utrecht, The Netherlands

Christiaans, H and Dorst, C (1992) Cognitive Models in Industrial Design Engineering: a protocol study, in D L Taylor and D A Stauffer (eds.), *Design Theory and Methodology – DTM92,* American Society of Mechanical Engineers, New York, USA

Cross, A (1980) Design and General Education *Design Studies* Vol 1 No 4 pp. 202–206

Cross, A (1984) Towards an Understanding of the Intrinsic Values of Design Education, *Design Studies* Vol 5, No 1, pp. 31–39

Cross, N (1967) Simulation of Computer Aided Design, *MSc Thesis*, UMIST, Manchester, UK

Cross, N (ed.) (1984) *Developments In Design Methodology* John Wiley and Sons Ltd., Chichester, UK

Cross, N (1989) *Engineering Design Methods: Strategies for Product Design* John Wiley and Sons Ltd., Chichester, UK

Cross, N (1999) Design Research: a disciplined conversation *Design Issues* Vol. 15, No. 2, pp. 5–10

Cross, N (2001) Achieving Pleasure From Purpose: the methods of Kenneth Grange, product designer *Design Journal* Vol 4, No 1, pp 48–58

Cross, N, Christiaans, H and Dorst, K (1994) Design Expertise Amongst Student Designers *Journal of Art and Design Education* Vol 13, No 1, pp. 39–56

Cross, N, Christiaans, H and Dorst, K (eds.) (1996) *Analysing Design Activity* John Wiley and Sons Ltd., Chichester, UK

Cross, N and Clayburn Cross, A (1995) Observations of Teamwork and Social Processes in Design *Design Studies* Vol. 16, No. 2, pp. 143–170

Cross, N and Clayburn Cross, A (1996) Winning by Design: the methods of Gordon Murray, racing car designer *Design Studies* Vol 17, No 1, pp. 91–107

Cross, N and Clayburn Cross, A (1998) Expertise in Engineering Design *Research in Engineering Design* Vol 10, No 3, pp. 141–149

Cross, N and Dorst, K (1998) Co-evolution of Problem and Solution Spaces in Creative Design: observations from an empirical study, in J Gero and M L Maher (eds.), *Computational Models of Creative Design IV,* University of Sydney, NSW, Australia

Cross, N, Dorst, K and Roozenburg, N (eds.) (1992) *Research in Design Thinking*, Delft University Press, Delft, The Netherlands

Cross, N and Nathenson, M (1981) Design Methods and Learning Methods, in J Powell and R Jacques (eds.), *Design:Science:Method*, Westbury House, Guildford, UK

Cross, N, Naughton, J and Walker, D (1981) Design Method and Scientific Method *Design Studies* Vol 2, No 4, pp. 195–201

Cross, N and Roozenburg, N (1992) Modelling the Design Process in Engineering and in Architecture *Journal of Engineering Design* Vol. 3, No. 4, pp. 325–337

Daley, J, Design Creativity and the Understanding of Objects, *Design Studies*, Vol 3, No 3, pp. 133–137

Darke, J (1979) The Primary Generator and the Design Process *Design Studies* Vol 1, No 1, pp 36–44

Davies, R (1985) A Psychological Enquiry into the Origination and Implementation of Ideas, *MSc Thesis*, UMIST, Manchester, UK

Davies, S and Castell, A (1992) Contextualizing Design: narratives and rationalization in empirical studies of software design *Design Studies* Vol 13, No 4, pp. 379–392

van Doesberg, T (1923) Towards a Collective Construction, *De Stijl* (Quoted by Naylor, G *The Bauhaus*, Studio Vista, London, 1968)

Dorst, K (1997) Describing Design: a comparison of paradigms, *PhD Thesis*, Faculty of Industrial Design Engineering, Delft University of Technology, The Netherlands

Dorst, K and Dijkhuis, J (1995) Comparing Paradigms for Describing Design Activity *Design Studies* Vol 16, No 2, pp. 261–274

Douglas, M and Isherwood, B (1979) *The World of Goods* Allen Lane, London, UK

Eastman, C M (1970) On the Analysis of Intuitive Design Processes, in G T Moore (ed.), *Emerging Methods in Environmental Design and Planning* MIT Press, Cambridge, MA, USA

Edwards, B (1979) *Drawing on the Right Side of the Brain* Tarcher, Los Angeles, CA, USA

Ericsson, K A and Lehmann, A (1996) Expert and Exceptional Performance: evidence on maximal adaptations on task constraints *Annual Review of Psychology* Vol 47, pp. 273–305

Ericsson, K A and Simon, H A (1993) *Protocol Analysis: Verbal Reports as Data* MIT Press, Cambridge, MA, USA

Ericsson, K A and Smith, J (eds.) (1991) *Toward a General Theory of Expertise: Prospects and Limits* Cambridge University Press, Cambridge, UK

Ferguson, E S (1977) The Mind's Eye: non-verbal thought in technology *Science* Vol 197, No 4306

Ferguson, E S (1992) *Engineering and the Mind's Eye*, MIT Press, Cambridge, MA, USA

Fox, R (1981) Design-based Studies: an action-based 'form of knowledge' for thinking, reasoning, and operating *Design Studies* Vol 2, No 1, pp. 33–40

Frankenberger, E and Badke-Schaub, P (1998) Integration of Group, Individual and External Influences in the Design Process, in E Frankenberger, P Badke-Schaub and H Birkhofer (eds.), *Designers – The Key to Successful Product Development,* Springer, London, UK

French, M J (1979) A Justification for Design Teaching in Schools *Engineering* (design education supplement) p. 25

French, M J (1985) *Conceptual Design for Engineers*, Design Council, London, UK

French, M J (1994) *Invention and Evolution: Design in Nature and Engineering* Cambridge University Press, Cambridge, UK

Fricke, G (1993) Empirical Investigations of Successful Approaches When Dealing With Differently Précised Design Problems *International Conference on Engineering Design ICED93*, Heurista, Zürich

Fricke, G (1996) Successful Individual Approaches in Engineering Design *Research in Engineering Design* Vol 8, pp. 151–165

Galle, P (1996) Design Rationalisation and the Logic of Design: a case study *Design Studies* Vol 17, No 3, pp. 253–275

Gardner, H (1983) *Frames of Mind: The Theory of Multiple Intelligences*, Heinemann, London, UK

Gasparski, W and Strzalecki, A (1990) Contributions to Design Science: praxeological perspective *Design Methods and Theories* Vol 24, No 2

Gazzaniga, M S (1970) *The Bisected Brain*, Appleton Century Crofts, New York, USA

Gelernter, D (1994) *The Muse in the Machine: Computers and Creative Thought* Fourth Estate, London, UK

Gero, J (1994) Computational Models of Creative Design Processes, in Dartnall, T. (ed.), *Artificial Intelligence and Creativity* Kluwer, Dordrecht, The Netherlands

Gero, J and McNeill, T (1998) An Approach to the Analysis of Design Protocols *Design Studies* Vol 19, No 1, pp. 21–61,

Glynn, S (1985) Science and Perception as Design *Design Studies* Vol 6, No 3, pp. 122–126

Goel, V (1995) *Sketches of Thought*, MIT Press, Cambridge, MA, USA,

Goel, V and Pirolli, P (1992) The Structure of Design Problem Spaces *Cognitive Science* Vol 16, pp. 395–429

Göker, M H (1997) The Effects of Experience During Design Problem Solving *Design Studies* Vol 18, No 4, pp. 405–426

Goldschmidt, G (1991) The Dialectics of Sketching *Creativity Research Journal* Vol 4, No 2, pp. 123–143

Goldschmidt, G (1996), The Designer as a Team of One, in Cross, N. *et al.* (eds.), *Analysing Design Activity*, John Wiley and Sons Ltd., Chichester, UK

Gordon, W J (1961) *Synectics: The Development of Creative Capacity*, Harper and Brothers, New York, NY, USA

Grant, D (1979) Design Methodology and Design Methods *Design Methods and Theories* Vol 13, No 1

Gregory, S A (1966a) Design and the Design Method, in Gregory, S A (ed.) *The Design Method* Butterworth, London, UK

Gregory, S A (1966b) Design Science, in Gregory, S A (ed.) *The Design Method*, Butterworth, London, UK

Guindon, R (1990a) Knowledge Exploited by Experts During Software System Design *International Journal of Man-Machine Studies* Vol 33, pp. 279–304

Guindon, R (1990b) Designing the Design Process: exploiting opportunistic thoughts *Human-Computer Interaction* Vol 5, pp. 305–344

Günther, J, Frankenberger, E and Auer, P (1996) Investigation of Individual and Team Design Processes in Mechanical Engineering, in Cross, N. *et al.* (eds.), *Analysing Design Activity*, John Wiley and Sons Ltd., Chichester, UK

Hansen, F (1974) *Konstruktionswissenschaft*, Carl Hanser, Munich, Germany

Harrison, A (1978) *Making and Thinking* Harvester Press, Hassocks, Sussex, UK

Hillier, B and Leaman, A (1974) How is Design Possible? *Journal of Architectural Research* Vol 3, No 1, pp. 4–11

Hillier, B and Leaman, A (1976) Architecture as a Discipline *Journal of Architectural Research* Vol 5, No 1, pp. 28–32

Holyoak, K J (1991) Symbolic Connectionism: toward third-generation theories of expertise, in Ericsson, K A and Smith, J (eds.), *Toward a General Theory of Expertise: Prospects and Limits* Cambridge University Press, Cambridge, UK

Hubka, V (1982) *Principles of Engineering Design*, Butterworth, Guildford, UK

Hubka, V and Eder, W E (1987) Scientific Approach to Engineering Design, *Design Studies* Vol 8, No 3, pp. 123–137

Jacques, R and Powell, J (eds.) (1981) *Design:Science:Method*, Westbury House, Guildford, UK

Jansson, D G and Smith, S M (1991) Design Fixation *Design Studies* Vol 12, No 1, pp. 3–11

Jones, J C (1970) *Design Methods* Wiley, Chichester, UK

Jones, J C (1977) How My Thoughts About Design Methods Have Changed During the Years, *Design Methods and Theories*, Vol 11, No 1, pp. 50–62

Jones, J C and Thornley, D G (eds.) (1963) *Conference on Design Methods*, Pergamon, Oxford, UK

Kavakli, M, Scrivener, S *et al.* (1998) Structure in Idea Sketching Behaviour *Design Studies* Vol 19, No 4, pp. 485–517

Koestler, A (1964) *The Act of Creation*, Hutchinson and Co. Ltd., London, UK

Kolodner, J L and Wills, L M (1996) Powers of Observation in Creative Design *Design Studies* Vol 17, No 4, pp. 385–416

Lakatos, I (1970) Falsification and the Methodology of Scientific Research Programmes, in Lakatos, I and Musgrave, A (eds.) *Criticism and the Growth of Knowledge* Cambridge University Press, Cambridge, UK

Lasdun, D (1965) An Architect's Approach to Architecture *RIBA Journal* Vol 72, No 4

Lawson, B (1979) Cognitive Strategies in Architectural Design *Ergonomics* Vol 22, No 1, pp. 59–68

Lawson, B (1980) *How Designers Think* Architectural Press, London, UK

Lawson, B (1994) *Design In Mind*, Butterworth-Heinemann, Oxford, UK

Le Corbusier (1929) *CIAM 2nd Congress*, Frankfurt, Germany

Levin, P H (1966) Decision Making in Urban Design *Building Research Station Note EN51/66* Building Research Station, Garston, Herts, UK

Lloyd, P, Lawson, B and Scott, P (1996) Can Concurrent Verbalisation Reveal Design Cognition? *Design Studies* Vol 16, No 2, pp. 237–259

Lloyd, P and Scott, P (1994) Discovering the Design Problem *Design Studies* Vol 15, No 2, pp. 125–140

Lloyd, P and Scott, P (1995) Difference in Similarity: interpreting the architectural design process *Planning and Design* Vol 22, pp. 383–406

Maccoby, M (1991) The Innovative Mind at Work *IEEE Spectrum* December, pp. 23–35

MacCormac, R (1976) *Design Is...* (Interview with N. Cross), BBC/Open University TV programme, BBC, London, UK

McGown, A, Green, G *et al.* (1998) Visible Ideas: information patterns of conceptual sketch activity *Design Studies* Vol 19, No 4, pp. 431–453

McNeill, T, Gero, J *et al.* (1998) Understanding Conceptual Electronic Design Using Protocol Analysis *Research in Engineering Design* Vol 10, No 3, pp. 129–140

McPeck, J E (1981) *Critical Thinking and Education* Martin Robertson, Oxford, UK

March, L J (1976) The Logic of Design and the Question of Value, in March, L J (ed.) *The Architecture of Form* Cambridge University Press, Cambridge, UK

Marples, D (1960) *The Decisions of Engineering Design* Institute of Engineering Designers, London, UK

Mazijoglou, M, Scrivener, S and Clark, S (1996) Representing Design Workspace Activity, in Cross, N. *et al.* (eds.), *Analysing Design Activity*, John Wiley and Sons Ltd., Chichester, UK

Ornstein, R E (1975) *The Psychology of Consciousness* Jonathan Cape, London; Penguin Books, Harmondsworth, UK

Pahl, G and Beitz, W (1984) *Engineering Design* Springer/Design Council, London, UK

Peters, R S (1965) Education as Initiation, in Archambault, R D (ed.) *Philosophical Analysis and Education* Routledge and Kegan Paul, London, UK

Pugh, S (1991) *Total Design: integrated methods for successful product engineering* Addison-Wesley, Wokingham, UK

Purcell, A T and Gero, J (1991) The Effects of Examples on the Results of Design Activity, in Gero, J S (ed.) *Artificial Intelligence in Design AID91* Butterworth-Heinemann, Oxford, UK

Purcell A T, Williams P, *et al.* (1993) Fixation Effects: do they exist in design problem solving? *Environment and Planning B: Planning and Design* Vol 20, No 3, pp. 333–345

Purcell, T and Gero, J (1996) Design and Other Types of Fixation *Design Studies* Vol 17, No 4, pp. 363–383

Pye, D (1978) *The Nature and Aesthetics of Design* Barrie and Jenkins, London, UK

Radcliffe, D (1996) Concurrency of Actions, Ideas and Knowledge Displays Within a Design Team, in Cross, N. *et al.* (eds.), *Analysing Design Activity*, John Wiley and Sons Ltd., Chichester, UK

Radcliffe, D and Lee, T Y (1989) Design Methods Used by Undergraduate Engineering Students *Design Studies* Vol 10, No 4, pp. 199–207

Rittel, H and Webber, M (1973) Dilemmas in a General Theory of Planning *Policy Sciences* Vol 4, pp. 155–169

Roozenberg, N (1993) On the Pattern of Reasoning in Innovative Design *Design Studies*, Vol 14, No 1, pp. 4–18

Rosenman, M and Gero, J (1993) Creativity in Design Using a Prototype Approach, in Gero, J and Maher, M L (eds.) *Modeling Creativity and*

Knowledge-Based Creative Design, Lawrence Erlbaum Associates, Hillsdale, New Jersey, USA

Rowe, P (1987) *Design Thinking*, MIT Press, Cambridge, MA, USA

Royal College of Art (1979) *Design in General Education* Department of Design Research, Royal College of Art, London, UK

Roy, R (1993) Case Studies of Creativity in Innovative Product Development, *Design Studies* Vol 14, No 4, pp. 423–443.

Ryle, G (1949) *The Concept of Mind* Hutchinson, London, UK

Schön, D (1983) *The Reflective Practitioner*, Temple-Smith, London, UK

Schön, D (1988) Designing: rules, types and worlds *Design Studies* Vol 9, No 3, pp. 181–190

Schön, D and Wiggins, G (1992) Kinds of Seeing and their Functions in Designing *Design Studies* Vol 13, No 2, pp. 135–156

Simon, H A (1969) *The Sciences of the Artificial* MIT Press, Cambridge, MA, USA

Smith, R P and Tjandra, P (1998) Experimental Observation of Iteration in Engineering Design *Research in Engineering Design* Vol 10, No 2, pp. 107–117

Sperry, R W, Gazzaniga, M S and Bogen, J E (1969) Interhemispheric Relations: the neocortical commissures; syndromes of hemispheric disconnection, in Vinken, P J and Bruyn, G W (eds.), *Handbook of Clinical Neurology*, Vol 4, North-Holland, Amsterdam, The Netherlands

Stauffer, L, Ullman, D *et al.* (1987) Protocol Analysis of Mechanical Engineering Design *International Conference on Engineering Design ICED87*, ASME, New York, USA

Suwa, M, Purcell, T and Gero, J (1998) Macroscopic Analysis of Design Processes Based on a Scheme for Coding Designers' Cognitive Actions *Design Studies* Vol 19, No 4, pp. 455–483

Suwa, M and Tversky, B (1997) What do Architects and Students Perceive in Their Design Sketches? *Design Studies* Vol 18, No 4, pp. 385–403

Takeda, H, Yoshioka M, *et al.* (1996) Analysis of Design Protocol by Functional Evolution Process Model, in Cross, N *et al.* (eds.), *Analysing Design Activity*, John Wiley and Sons Ltd., Chichester, UK

Thomas, J C and Carroll, J M (1979) The Psychological Study of Design *Design Studies* Vol 1, No 1, pp. 5–11

Tjalve, E (1979) *A Short Course in Industrial Design*, Newnes-Butterworth, London, UK

Tzonis, A (1992) *Invention Through Analogy: lines of vision, lines of fire* Faculty of Architecture, Delft University of Technology, Delft, The Netherlands

Ullman, D G, Dietterich, T G *et al.* (1988) A Model of the Mechanical Design Process Based on Empirical Data *A I in Engineering Design and Manufacturing* Vol 2, No 1, pp. 33–52

Ullman, D G, Wood, S, *et al.* (1990) The Importance of Drawing in the Mechanical Design Process *Computers and Graphics* Vol 14, No 2, pp. 263–274

Valkenburg, R and Dorst, K (1998) The Reflective Practice of Design Teams *Design Studies* Vol 19, No 3, pp. 249–272

VDI (Verein Deutscher Ingenieure) (1987) *Design Guideline 2221: Systematic Approach to the Design of Technical Systems and Products* (English translation of 1985 German edition), VDI Verlag, Düsseldorf, Germany

Verstijnen, I M, Hennessey, J M, *et al.* (1998) Sketching and Creative Discovery *Design Studies* Vol 19, No 4, pp. 519–546

Visser W (1990) More or Less Following a Plan During Design: opportunistic deviations in specification *International Journal of Man-Machine Studies* Vol 33, pp. 247–278

de Vries, M, Cross, N and Grant, D (eds.) (1993) *Design Methodology and Relationships with Science*, Kluwer, Dordrecht, The Netherlands

Waldron, M B, and Waldron, K J (1988) A Time Sequence Study of a Complex Mechanical System Design *Design Studies* Vol 9, No 2, pp. 95–106

Wallas, G (1926) *The Art of Thought* Jonathan Cape, London, UK

Whitehead, A N (1932) Technical Education and its Relation to Science and Literature, in Whitehead, A N *The Aims of Education* Williams and Norgate, London, UK

Willem, R A (1990) Design and Science *Design Studies* Vol 11, No 1, pp. 43–47

Index

LaVergne, TN USA
03 May 2010
181385LV00001B/92/P